全国科学技术名词审定委员会

科学技术名词·工程技术卷（全藏版）

12

海峡两岸船舶工程名词

海峡两岸船舶工程名词工作委员会

国家自然科学基金资助项目

科 学 出 版 社

北 京

内 容 简 介

　　本书是全国科学技术名词审定委员会、台湾中华船舶机械工程学会共同组织海峡两岸船舶工程界专家会审的海峡两岸船舶工程名词对照本，是在海峡两岸各自正式出版的船舶工程名词的基础上选择船舶专用名词对照编订而成。内容包括船舶性能、船体结构、船舶机械、船舶电气、船舶通信与导航、舾装、船舶工艺等，共收词 3200 条。供海峡两岸船舶工程界和相关领域的人士使用。

图书在版编目(CIP)数据

科学技术名词. 工程技术卷：全藏版 / 全国科学技术名词审定委员会审定.
—北京：科学出版社，2016.01
　ISBN 978-7-03-046873-4

　I. ①科⋯　II. ①全⋯　III. ①科学技术–名词术语 ②工程技术–名词术语
　IV. ①N-61 ②TB-61

　中国版本图书馆 CIP 数据核字(2015)第 307218 号

责任编辑：李玉英 / 责任校对：陈玉凤
责任印制：张　伟 / 封面设计：铭轩堂

科 学 出 版 社 出版
北京东黄城根北街 16 号
邮政编码：100717
http://www.sciencep.com
北京厚诚则铭印刷科技有限公司印刷
科学出版社发行　各地新华书店经销
*
2016 年 1 月第 一 版　　开本：787×1092 1/16
2016 年 1 月第一次印刷　　印张：13 3/4
字数：326 000
定价：7800.00 元(全 44 册)
(如有印装质量问题，我社负责调换)

海峡两岸船舶工程名词工作委员会委员名单

召 集 人：孙 英

委　　员（按姓氏笔画为序）：

王荣生　　王勉钰　　朱遐良　　李 强　　李玉英

李贵臣　　金向军　　周光霁　　赵 华　　程天柱

召 集 人：王偉輝

委　　員（按姓氏筆畫爲序）：

朱于益　　李中岳　　邱逢琛　　辛敬業　　吳重雄

林益煌　　林輝政　　胡平祥　　洪振發　　莊政義

陸磐安　　陳定雄　　陳建宏　　黃正利　　黃正清

張達禮　　鄭正村　　鄭勝文

序

　　科学技术名词作为科技交流和知识传播的载体,在科技发展和社会进步中起着重要作用。规范和统一科技名词,对于一个国家的科技发展和文化传承是一项重要的基础性工作和长期性任务,是实现科技现代化的一项基础性工程。没有这样一个系统的规范化的基础条件,不仅现代科技的协调发展将遇到困难,而且,在科技广泛渗入人们生活各个方面、各个环节的今天,还将会给教育、传播、交流等方面带来困难。

　　科技名词浩如烟海,门类繁多,规范和统一科技名词是一项十分繁复和困难的工作,而海峡两岸的科技名词要想取得一致更需两岸同仁作出坚韧不拔的努力。由于历史的原因,海峡两岸分隔逾50年。这期间正是现代科技大发展时期,两岸对于科技新名词各自按照自己的理解和方式定名,因此,科技名词,尤其是新兴学科的名词,海峡两岸存在着比较严重的不一致。同文同种,却一国两词,一物多名。这里称"软件",那里叫"软体";这里称"导弹",那里叫"飞弹";这里写空间,那里写太空;如果这些还可以沟通的话,这里称"等离子体",那里称"电浆";这里称"信息",那里称"资讯",相互间就不知所云而难以交流了。"一国两词"较之"一国两字"造成的后果更为严峻。'一国两字'无非是两岸有用简体字的,有用繁体字的,但读音是一样的,看不懂,还可以听懂。而"一国两词"、"一物多名"就使对方既看不明白,也听不懂了。台湾清华大学的一位教授前几年曾给时任中科院院长周光召院士写过一封信,信中说:1993年底两岸电子显微学专家在台北举办两岸电子显微学研讨会,会上两岸专家是以台湾国语、大陆普通话和英语三种语言进行的。这说明两岸在汉语科技名词上存在着差异和障碍,不得不借助英语来判断对方所说的概念。这种状况已经影响两岸科技、经贸、文教方面的交流和发展。

　　海峡两岸各界对两岸名词不一致所造成的语言障碍有着深刻的认识和感受。具有历史意义的"汪辜会谈"把探讨海峡两岸科技名词的统一列入了共同协议之中,此举顺应两岸民意,尤其反映了科技界的愿望。两岸科技名词要取得统一,首先是需要了解对方。而了解对方的一种好的方式就是编订名词对照本,在编订过程中以及编订后,经过多次的研讨,逐步取得一致。

　　全国科学技术名词审定委员会(简称全国科技名词委)根据自己的宗旨和任务,始终把海峡两岸科技名词的对照统一工作作为责无旁贷的历史性任务。近些年一直本着积极推进,增进了解;择优选用,统一为上;求同存异,逐步一致的精神来开展这项工作。先后接待和安排了许多台湾同仁来访,也组织了多批专家赴台参加有关学科的名词对照研讨会。工作中,按照先急后缓、先易后难的精神来安排。对于那些与"三通"

有关的学科,以及名词混乱现象严重的学科和条件成熟、容易开展的学科先行开展名词对照。

在两岸科技名词对照统一工作中,全国科技名词委采取了"老词老办法,新词新办法",即对于两岸已各自公布、约定俗成的科技名词以对照为主,逐步取得统一,编订两岸名词对照本即属此例。而对于新产生的名词,则争取及早在协商的基础上共同定名,避免以后再行对照。例如101~109号元素,从9个元素的定名到9个汉字的创造,都是在两岸专家的及时沟通、协商的基础上达成共识和一致,两岸同时分别公布的。这是两岸科技名词统一工作的一个很好的范例。

海峡两岸科技名词对照统一是项长期的工作,只要我们坚持不懈地开展下去,两岸的科技名词必将能够逐步取得一致。这项工作对两岸的科技、经贸、文教的交流与发展,对中华民族的团结和兴旺,对祖国的和平统一与繁荣富强有着不可替代的价值和意义。这里,我代表全国科技名词委,向所有参与这项工作的专家们致以崇高的敬意和衷心的感谢!

值此两岸科技名词对照本问世之际,写了以上这些,权当作序。

2002 年 3 月 6 日

前　　言

　　造船产业是世界性的产业,现在两岸的年造船总吨位已经跃居世界第三。在海峡两岸同时加入世界贸易组织(WTO)之时,将两岸船舶工程名词进行对照,便于两岸造船界对科学技术名词的理解和交流,促进两岸船舶工业的互相联系和贸易,这是海峡两岸船舶工程技术人员的共识。自1996年起,两岸船舶工程方面的专家开始了船舶工程名词对照的准备工作。

　　1996年初,中国造船工程学会学术代表团访台期间,将基本定稿的《船舶工程名词》赠予台湾船舶工程专家,以加强两岸造船界的交流与合作。台湾中华船舶机械工程学会对此项工作非常积极和重视,召集专家成立工作小组,以"国立编译馆"1963年出版的《造船工程名词》为依据,参考大陆提供的《船舶工程名词》出版稿,编制了《两岸造船工程名词对照》。1998年在全国科学技术名词审定委员会代表团访台期间,台湾方面提交了《两岸造船工程名词对照》,并建议大陆专家校核。

　　2000年9月,全国科学技术名词审定委员会向中国造船工程学会通报了台湾中华船舶机械工程学会关于加强两岸船舶工程名词术语交流与合作的意向。2001年初,中国造船工程学会成立了以中国船舶工业综合技术经济研究院专家为主体的海峡两岸船舶工程名词工作委员会,开展以中国标准出版社1996年出版的《舰船标准术语词典　通用术语》(收录术语13 010条)为依据,以台湾中华船舶机械工程学会最新提供的《造船工程名词》为参考的对照工作。

　　经过两岸船舶工程技术专家的共同努力,于2001年8月完成了《海峡两岸船舶工程名词》初稿,收录名词12 985条。2001年11月由全国科学技术名词审定委员会和中国造船工程学会组成的大陆船舶工程名词工作委员会9名专家赴台参加两岸船舶工程名词对照第一次研讨会,台湾造船工程界有13名专家参加了会议。会议气氛热烈、同行态度严谨,两岸代表就会议取得的共识签署了会议纪要。会议明确成立海峡两岸船舶工程名词工作委员会,两岸委员总计30人,并就对照的任务、收词范围、基本原则、工作方式、文稿格式及工作规划等形成了一致意见。

　　会后,两岸专家进一步核对、修订,完成第二次研讨会的讨论稿。经两岸专家校核,选定5 015条名词进行对照,删去了大量的通用名词。

　　2002年7月,两岸专家在昆明举行了海峡两岸船舶工程名词第二次研讨会。会上对选出的600余条名词重点进行了研讨。会议在融洽、热烈的气氛中通过了会议纪要,确定了编辑、出版等事宜。

　　第二次研讨会确定涵义不清或通用性名词不收录,因此两岸专家对讨论稿进行了再次的删减。

从 2002 年 9 月底至 2003 年 4 月,经过两岸专家的反复推敲修订《海峡两岸船舶工程名词》终于定稿,全书共收录对照的名词 3 200 条。

出版《海峡两岸船舶工程名词》,是一件承前启后、福荫子孙的好事,也是海峡两岸造船界的一件喜事。我们有幸完成此项工作,感到荣幸之至。《海峡两岸船舶工程名词》凝结了两岸船舶工程技术专家的心血,是我们共同的智慧结晶。然而,由于我们的学识、能力有限,其中的疏漏在所难免,还望同仁们指正。

海峡两岸船舶工程名词对照是一项任重而道远的工作,随着世界造船技术的不断发展,两岸造船界还有待于进一步深化研究,不断补充新的船舶工程名词,各自修改不当的名词,及时进行交流与合作,力求船舶工程名词的逐步统一。

海峡两岸船舶工程名词工作委员会
2003 年 5 月

编 排 说 明

一、本书是海峡两岸船舶工程名词对照本。

二、本书分正篇和副篇两部分。正篇按汉语拼音顺序编排;副篇按英文名的字母顺序编排。

三、[ ]中的字使用时可以省略。

正篇

四、正名和异名分别排序,在异名处用(=)注明正名和英文名同义词。

五、对应的英文名为多个时(包括缩写词)用","分隔。

副篇

六、英文名所对应的相同概念的汉文名用","分隔,不同概念的用① ② ③分别注明。

七、英文名的同义词或近义词用(=)注明。

八、英文缩写词排在全称后的( )内。

目　　录

正 篇

A

祖 国 大 陆 名	台 湾 地 区 名	英 文 名
安全灯	安全燈	safe-light, safety lamp
安全阀	安全閥	safety valve
安全开关	安全開關	safety, safety switch
安全索	救生索，攀手索	life line
鞍形舱	鞍形艙	saddle chamber
岸电电缆	岸電電纜	shore connection cables
岸电箱	岸電接線盒	shore connection box

B

祖 国 大 陆 名	台 湾 地 区 名	英 文 名
巴拿马运河导缆孔	巴拿馬運河導索器	Panama chock
巴拿马运河吨位	巴拿馬運河噸位	Panama canal tonnage
巴拿马运河吨位证书	巴拿馬運河噸位證書	Panama Canal Tonnage Certificate
巴氏合金	巴比合金，白合金	Babbitt metal
靶船	靶船	target craft, target ship
白昼信号灯	日間訊號燈	daylight signalling light
摆墩	擺墩，塢墩佈置	blocking arrangement
板缝排列		seam arrangement
板料	金屬薄板，板金	sheet metal
板列	板列，列板	panel, strake
半闭式给水系统	半閉式給水系統	semi-closed feed water system
半梁	半梁	half beam
半潜船	半潛式船	semi-submerged ship
半潜式钻井平台	半潛式鑽探平台	semi-submersible drilling unit, semi-submersible drilling platform
半坞式船台	半塢式船台	semi-dock building berth
半悬舵	半懸舵，半平衡舵	partially underhung rudder, semi-balanced

·1·

祖 国 大 陆 名	台 湾 地 区 名	英 文 名
		rudder
伴流	伴流,跡流	wake
伴流测量	跡流量测	wake measurement, wake survey
伴流分数	跡流因數,跡流係數	wake fraction, wake coefficient
伴流模拟	跡流模擬	wake simulation
伴流因数	跡流因數,跡流係數	wake factor
邦戎曲线	龐琴曲線	Bonjean's curves
棒料	棒料	bar
包覆	包覆,被覆	covering, cladding
包装	包裝	package, packing
包装舱容	包裝貨容積	bale cargo capacity
包装件	包裝件	pack, package
保护电流密度	保護電流密度	protection current density
保护电位	保護電位	protective potential
保护膜	膜片	diaphragm
保护性覆盖层	保護性塗層	protective coating
保温层	[隔熱]襯套	lagging
保温用具	保溫具	thermal protective aid
保险螺母	並緊螺帽	lock nut
报警装置	警報裝置	alarm unit
报务室	無線電室	radio room
爆炸焊	爆炸焊接	explosion welding
背空化	[葉]背空蝕	back cavitation
背压	背壓,反壓	back pressure
背压式汽轮机	背壓渦輪機	back-pressure turbine
背压调节器	背壓調整器	back pressure regulator
备用泵	備用泵	stand by pump
备用发电机组	備用發電機組	stand-by generating set
倍频程	倍頻帶	octave
本质安全电路	本質安全線路	intrinsically safe circuit
泵舱	泵室	pump room
比功[率]	比馬力,比功率	specific power
比例放样		scale lofting
舭	舭	bilge
舭部半径	舭曲半徑	bilge radius
舭部扶手	舭部扶手	underside handholds
舭部升高	舭横斜高	deadrise
舭墩	舭邊墩	bilge block

祖国大陆名	台湾地区名	英 文 名
舭列板	舭板列	bilge strake
舭龙骨	舭龍骨	bilge keel, rolling chock
舭肘板	舭腋板	bilge bracket
避雷器	避雷器	surge arrester, lightning arrester
避碰灯		anti-collision light
避碰装置		equipment for collision avoidance
边舱	翼櫃,翼艙	wing tank
边墩	邊墩	side keel block
边缘加工	邊緣[預]加工	edge preparation
变幅绞车	跨索絞機	span winch, spanwire winch
变流机	換流機	convertor
变螺距	可變節距,可變螺距	variable pitch
变频器	變頻器,變頻機	frequency convertor, frequency changer
变相机	換相機	phase convertor
变形	變形	deformation
标准排放接头	標準排泄接頭	standard discharge connection
标准排水量	標準排水量(軍艦)	standard displacement
表面波	表面波	surface wave
表面处理	表面處理	surface treatment
表面粗糙度	表面粗糙度	surface roughness
表面力	表面力	surface force
表面式回热器	複熱器	recuperator
表面式凝汽器	表面冷凝器	surface condenser
表面预处理	表面預處理	surface pretreatment
冰舱	冰艙	ice bunker, ice hold
冰带区	冰帶(外板),冰帶板列	ice belt
冰锚	冰錨	ice anchor
冰载荷	冰負荷	ice load
并联运行	並聯運轉(電)	parallel operation
并列断续角焊缝	並列斷續填角焊接	chain intermittent fillet weld
波	波	wave
波长	波長	wave length
波腹	波腹	antinode, wave loop
波激振动	船體波振	springing
波浪谱	波譜	wave spectrum
波浪水动压力修正	史密斯修正波效應	Smith correction
波浪载荷	波浪負荷	wave load
波浪中阻力试验	波浪中阻力試驗	resistance test in waves

祖 国 大 陆 名	台 湾 地 区 名	英 文 名
波列	波列	wave train
波前	波前[進面]	wave front
波倾角	波面斜率,波面傾角	slope of wave surface
波形	波形	wave form
波形舱口盖	波形艙蓋	corrugated hatchcover
波型阻力	波型阻力	wave pattern resistance
玻璃钢船	玻[璃]纖[維]強化塑膠船	fiberglass reinforced plastic ship, FRP ship, fiberglass reinforced plastic boat
玻璃纤维增强塑料	玻[璃]纖[維]強化塑膠	glass fiber reinforced plastics, GRP, fiberglass reinforced plastics, FRP
驳船	駁船	barge, lighter
泊位	泊位	berth
补偿电线	補償線圈	compensating wire, compensating winding
补偿器	補償器	compensator
补给水	補充給水	make-up feed water
补焊	補焊	repair welding
补水储存舱	備用給水櫃	reserve feedwater tank
捕鲸船	捕鯨船	whaler
捕鲸母船	鯨加工船	whale factory ship
捕捞机械	漁撈機械	fishing machinery
不沉性	不沉性	insubmersibility
不定常空泡	不穩定空泡	unsteady cavities
不对称浸水	不對稱浸水	unsymmetrical flooding
不连续[性]	不連續性	discontinuity
不平衡舵	不平衡舵	unbalanced rudder
不燃材料	不燃材料	non-combustible material
不稳定环	遲滯回圈,磁滯回圈	hysteresis loop
不稳定空泡	不穩定空泡	non-stationary cavities
布缆船	布纜船	cable layer, cable ship
布雷舰艇	佈雷艦,佈雷艇	minelayer
布网船	布網船	net layer
部分封闭救生艇	部分圍蔽救生艇	partially enclosed lifeboat
部分负荷	部分負荷	partial load

C

祖国大陆名	台湾地区名	英文名
采矿船		mining dredger
采珍渔船	采珠船	pearl boat, lugger
残油类标准排放接头	殘油排泄標準接頭	residual oil standard discharge connection
残余应力	殘留應力	residual stress, internal stress
舱	艙	compartment, hold
舱壁	艙壁	bulkhead
舱壁板	艙壁板	bulkhead plate
舱壁扶强材	艙壁防撓材	bulkhead stiffener
舱壁甲板	艙壁甲板	bulkhead deck
舱壁龛	艙壁凹入部	bulkhead recess
舱壁门	艙壁門	bulkhead door
舱底泵	舭〔水〕泵	bilge pump
舱底水系统	舭水系統	bilge system
舱底水总管	舭水總管	bilge main line, bilge main
舱底污水	舭水,艙底水	bilge water
舱口	艙口	hatch, hatchway, hatch opening
舱口端梁	艙口端梁	hatch end beam
舱口盖	艙口蓋	hatchcover
舱口盖绞车	艙口蓋絞機	hatchcover winch
舱口活动横梁	艙口活動梁	portable hatch beam, hatch beam
舱口围板	艙口緣圍	hatch coaming
舱口悬臂梁	艙口側半梁	hatch side cantilever beam, fork beam
舱口纵桁	艙口側縱梁	hatch side girder
舱面货	艙面貨,甲板貨	deck cargo
舱面属具	甲板裝具	deck equipment and fittings, deck fittings
舱容图	容積圖	capacity plan
舱室	艙室	space
舱室布置		interior arrangement
舱室设备		accommodation equipment
舱室通风机	艙室〔通〕風機	cabin fan
舱室属具		cabin outfit
操舵拉杆	操舵桿	steering rod
操舵链	舵鏈	steering chain

祖 国 大 陆 名	台 湾 地 区 名	英 文 名
操舵轮	舵輪	steering wheel
操舵目标灯	拖航燈	steering light
操舵索	操舵鋼索	steering wire
操舵台	操舵台	steering stand
操舵轴	操舵軸	steering shafting, steering shaft
操舵装置	操舵裝置	steering gear, steering apparatus
操纵限制灯	操縱限制燈	restricted maneuver light
操纵信号灯	操船信號燈	maneuvering light
操纵性衡准	操縱性準則	criteria of maneuverability
操纵性试验	操縱性試驗	maneuverability test
槽型舱壁	波形艙壁	corrugated bulkhead
侧壁气垫船	側壁式氣墊船	sidewall hovercraft, side waller
侧开泥驳	側漏斗型駁船	side hopper barge
侧面角焊缝	側面填角焊接	fillet weld in parallel shear
侧推器	推力裝置	thruster
侧向推力装置	側推裝置	side thrust device
侧斜	[螺]葉歪斜	skew back
侧斜角	[螺葉]歪斜角度	skew angle
侧移式舱口盖	側移式艙口蓋	side rolling hatchcover
测厚仪	厚度規	thickness gauge
测深	測深	depth sounder, sounding
测深尺	測深標尺	sounding rod
测深锤	測[深]錘,測深[鉛]錘	hand lead, sounding lead, sound lead
测深管	測深管	sounding pipe
测试段	測試段	measuring section
层间剪切强度	層間剪切強度	interlaminar shear strength
层状腐蚀	層狀腐蝕	layer corrosion
差动舵		Jenckel rudder
插拔桩状态	插拔樁狀態	spud driving and pulling condition
拆船	拆船	shipbreaking
拆验	拆檢	open-up examination, examination of opened up parts
柴-燃联合动力装置	柴油燃氣渦輪組合機	combined diesel and/or gas turbine power plant, CODAG, CODOG
柴油沉淀柜	柴油沉澱櫃	diesel oil settling tank
柴油发电机	柴油發電機	diesel generator, diesel dynamo
柴油发电机组	柴油發電機組	diesel generating set
柴油机	柴油引擎,狄塞爾引擎	diesel engine

祖 国 大 陆 名	台 湾 地 区 名	英 文 名
柴油机船	柴油機船,内燃機船	diesel ship, motor ship, diesel boat
柴油机电力推进装置		diesel-electric propulsion plant
柴油机动力装置		diesel engine power plant
柴油机通气管工作装置	［潛艇］通氣管裝置	snorkel, snort
柴油日用柜	柴油日用櫃	diesel oil daily tank
铲斗	杓	dipper
铲斗转盘	鏟斗轉盤	turntable of dipper machine
铲扬机	鏟揚機	dipper machine
长峰波	長峰波	long crested waves
长宽比	長寬比	length breadth ratio
长深比	長深比	length depth ratio
常规潜艇	傳統潛艇	conventional submarine
超短波通信		ultra short wave communication
超负荷	過［量］負荷	overload
超负荷试验	過負荷試驗	overload test
超高压锅炉	超高壓鍋爐	super high pressure boiler
超空化螺旋桨	超空化螺槳	super-cavitating propeller
超扭保护装置	超扭［矩］保護設施	overtorque protection device
超速	超速	overspeed
超速保护装置	超速保護設施	overspeed protection device
超越角	過沖角	overshoot angle
车间底漆	防銹底漆	shop primer
车辆舱	車輛艙	vehicle hold
车辆甲板	車輛甲板	wagon deck, vehicle deck, car deck
掣链钩	吊鏈鈎,拉線爪	devil's claw
掣链器	制鏈器(錨),錨鏈扣	chain stopper, deck stopper
掣锚器	止錨器	anchor stopper
掣索器	制索器	rope stopper
沉积速度	澱積率,堆積率	deposition rate
沉垫自升式钻井平台		mat jack-up drilling unit, mat jack-up drilling platform
沉雷(水雷)	沉雷	sunken mine
沉箱	沉箱,潛水箱	caisson
衬垫焊	襯墊焊法	welding with backing
衬里	内襯,襯料,襯層	lining
撑材	支桿,支柱	strut
撑杆	撐桿	bracing
成品油船	油品船	product carrier

祖国大陆名	台湾地区名	英 文 名
成形法	成型法	forming
承载能力	負荷能量	load carrying capacity
吃水	吃水	draft, draught
吃水标志	吃水標	draft mark, draught mark
吃水限制灯	吃水限制號燈	deep draught vessel light
持续常用功率	額定連續常用出力	continuous service rating, CSR
尺度效应	尺度效應	scale effect
充放电板	充放電盤	battery charging and discharging panel, charging and discharging board
充气式设备	已充氣救生設備	inflated appliance
充气试验	充氣試驗	air charging test, air test
冲荡	抖動	whipping
冲动式汽轮机	衝動渦輪機	impulse turbine
冲击	衝擊	impact
冲击腐蚀	衝擊腐蝕	impingement corrosion
冲孔	沖孔,穿孔	punching, piercing
冲水试验	射水試驗	hose test
冲翼艇	沖翼船	ram-wing craft
重叠船模	對叠[船]模,重叠模	double model
重叠螺旋桨	重叠型螺槳	overlap propeller, overlapping propeller
重热系数	再熱因數	reheat factor
抽舱排泥装置	抽艙排泥裝置	self-emptying installation
抽气	抽氣	bleed
抽气器	空氣抽射器	air ejector
抽汽式汽轮机	抽汽渦輪機	extraction turbine
出坞	出塢	undocking, undock
初检查	初次檢驗	original inspection, initial survey
初稳性	初穩度	initial stability
储备浮力	預[留]浮力	reserve buoyancy
储备排水量	排水量餘裕	displacement margin
储油船	儲油船	oil storage tanker
穿梭油船	穿梭油輪,短程往返油輪	shuttle tanker
传感器	感測器	sensor
船[舶]	船,艦	ship, vessel
船舶操纵性	船舶操縱性	ship maneuverability
船舶电气设备	船舶電氣設備	marine electrical equipment
船舶电站		marine electrical power plant

祖国大陆名	台湾地区名	英　文　名
船舶电站自动控制装置		automatic control system for marine electric power plant
船舶动力装置		marine power plant
船舶辅机	船舶輔機	marine auxiliary machinery
船舶改装	船舶改裝	ship conversion
船舶工程		ship engineering, maritime engineering
船舶国籍证书	船舶國籍證書	Certificate of Ship's Nationality
船舶快速性	船舶阻力性能	ship resistance and performance
船舶入级与建造规范	船舶入級與建造規範	rules for classification and construction of ships
船舶涂料	船舶塗料	marine paint
船舶推进	船舶推進	ship propulsion
船舶系统		ship system
船舶消防装置		marine fire fighting system
船舶摇荡	船舶搖蕩	ship oscillation
船舶噪声	船舶噪音	ship noise
船舶制冷装置		marine refrigerating plant
船舶阻力	船舶阻力	ship resistance
船侧骨架	側肋骨	side framing, side frame
船长	船長	length
船底板	[鋼]船底外板	bottom plating
船底横骨	[船]底肋骨	bottom frame
船底横桁	船底橫材	bottom transverse
船底涂料	船底塗料,船底漆	ship bottom paint, ship bottom coating
船底纵骨	[船]底縱肋	bottom longitudinal
船后螺旋桨试验	[螺槳]船後試驗	behind ship test of propeller, behind test
船级	船級	class of ship
船级社	船級協會,船級機構	classification society, register of shipping
船籍港	船籍港	port of registry
船模	船模	ship model, block model
船模阻力仪	阻力儀	resistance dynamometer
船身效率	船殼效率	hull efficiency
船台	造船台	building berth
船台标杆线	船台標竿線	vertical template for hull assembly
船台坡度	船台斜度,造船台坡度	slope of building berth, slope of ways, inclination of building slip
船台舾装		berth outfitting
船台装配		berth assembly

祖国大陆名	台湾地区名	英文名
船体	船體,船身,船殼	hull
[船体]放样	放樣	lofting
船体刚度	船體剛性	hull stiffness
船体固有振动频率	船體自然頻率	hull natural frequency
船体回路系统	船體回路系統	hull-return system
船体加工		hull steel fabrication
船体建造工艺		technology of hull construction
船体建造公差		tolerance of hull construction
船体结构	船體結構	hull structure, hull construction
[船体零件]号料		marking of hull parts
船体挠度	船體撓度	hull deflection
船体扭转振动	船體扭轉振動	hull torsional vibration
船体强度	船體強度	hull strength
船体线型	船形	hull form
船体型表面	船[體]型面	molded hull surface
船体振动	船體振動	ship vibration
船体装配		hull assembly
船坞	塢	dock
船坞舾装		dock outfitting
船舷接管	船舷接管	ship side pipe
船型系数	線型係數	coefficients of form
船用泵	船用泵	marine pump
船用变压器	船用變壓器	marine transformer
船用柴油机	船用柴油機	marine diesel engine
船用厨房设备	船用廚房設備	marine galley equipment
船用电力电缆	船用電力電纜	shipboard power cables
船用阀门	船用閥門	marine valves
船用防火板	船用防火板	marine fire proof panel
船用焚烧炉	船用焚化爐	marine incinerator
船用风机	船用風機	marine-type fan
船用广播设备	船用廣播系統	marine public address system
船用锅炉	船用鍋爐	marine boiler
船用焊条	船用焊條	shipbuilding electrode
船用火灾自动报警装置		marine automatic fire alarm system
船用家具	船用家具	marine furniture
船用胶合板	船用合板	marine plywood
船用内燃机	船用引擎	marine engine
船用气瓶		marine air bottle

祖 国 大 陆 名	台 湾 地 区 名	英 文 名
船用汽轮机	船用[蒸]汽[渦]輪機	marine steam turbine
船用燃气轮机	船用燃氣渦輪機	marine gas turbine
船用射频电缆	船用射頻電纜	shipboard radio-frequency cables
船用生活污水处理装置	船用[生活]污水處理系統	marine sewage treatment system
船用通信电缆	船用通信電纜	shipboard telecommunication cables
船用卫生设备	船用衛生設施	marine sanitary fixtures, marine sanitation device, MSD
船用蓄电池组		marine accumulator batteries, marine storage batteries
船用液压气动组件		marine hydropneumatic components and units
船员室	船員室	crew room, crew space
船长室	船長室	captain room, captain's room
船装	船體舾裝	hull outfitting, hull outfit
喘振	波振	surge
喘振线	顫動線,喘振線	surge line
串联式船坞		tandem dock
串列螺旋桨	重叠螺槳	tandem propeller
吹除阀	海底門清除閥	sea chest cleaning valve
吹灰器	吹灰器	soot blower
吹泥船		barge unloading dredger
垂荡	[船身]起伏	heaving
垂线间长	垂標間距	length between perpendiculars
垂向棱形系数	垂向棱塊係數	vertical prismatic coefficient
垂向弯曲振动	垂向彎曲振動	vertical flexural vibration
垂直式阻力动力仪	垂直式阻力動力計	vertical-type resistance dynamometer
捶击	拍底	pounding
磁场	磁場	magnetic field
磁粉检测	磁粉探傷檢查	magnetic particle testing
磁极	磁極	magnetic pole
磁流体动力推进装置	磁流體動力推進裝置	magneto hydrodynamic propulsion plant
磁罗经	磁羅經	magnetic compass
磁偏吹	磁吹	magnetic blow
磁通	磁通[量]	magnetic flux
磁致伸缩换能器	磁致伸縮換能器	magnetostrictive transducer
磁滞	磁滯	hysteresis
磁阻	磁阻	reluctance

祖国大陆名	台湾地区名	英 文 名
次要构件	次[要]構件	secondary member
刺网渔船	刺網漁船	gill netter
粗糙度阻力	粗面阻力	roughness resistance
粗滴过渡	粒狀傳遞	globular transfer
窜气	漏氣	blow-by
脆性断裂	脆性破裂	brittle fracture
淬[火]冷[却]	急冷,淬火	quenching
淬火冷却开裂	淬火裂痕	quench cracking, quenching crack
淬火冷却起始温度	淬火溫度	quenching temperature
淬火冷却应力	淬火應力	quenching stresses
淬透性	可硬化性	hardenability
淬硬性	硬化性,淬火性	hardening capacity
错移	偏位	offset
错油门	導閥	pilot valve

D

祖国大陆名	台湾地区名	英 文 名
搭接焊	搭焊	lap welding, lap weld
搭接接头	搭接	lap joint
搭接结点	搭接接頭	overlapping joint
打底焊道	打底焊接	backing run, backing weld
打捞船	救難船	salvage ship
打捞浮筒	打撈浮筒	lifting pontoon
打桩船	打樁船	floating pile driver, floating pile-driving plant
大气冷凝器	大氣冷凝器,大氣凝結器	atmospheric condenser
大气压	大氣壓力	atmospheric pressure
大抓力锚	高抓著力錨	high holding power anchor
带板	系板	band plate
带缆羊角	帶纜羊角	mooring cleat
带缆桩	系纜樁	bollard
带泥舱挖泥船	斗式挖泥船	hopper dredger
带式输送机	帶式運送機	belt conveyor
单道焊	單道焊接	single-pass welding
单底	單底	single bottom
单点系泊装置	單點系泊裝置	single point mooring unit

祖 国 大 陆 名	台 湾 地 区 名	英 文 名
单缸[内燃]机	單缸機	single-cylinder engine
单甲板船	單層甲板船	single-decked ship, single-decked vessel
单锚腿储油装置		single anchor leg storage, SALS
单锚腿系泊装置	單腳錨泊	single anchor leg mooring, SALM
单人常压潜水服	大氣壓潜水衣	atmospheric diving suit
单体船	單體船	mono-hull ship
单元舾装		unit outfitting
单元组装	單元組合,小組合	unit assembling
单轴系		single shafting
单作用泵	單動泵	single acting pump
单作用锤	落錘	drop hammer
淡水泵	淡水泵	fresh water pump
淡水舱	淡水艙,淡水櫃	fresh water tank
淡水加热器	淡水加熱器	fresh water heater
淡水冷却器	淡水冷卻器	fresh water cooler
淡水滤器	淡水過濾器	fresh water filter
弹钩	滑鈎	slip hook
弹药转运间	彈藥搬運室	ammunition handling room, ammunition carrier room
挡浪板	擋浪板	breakwater
导板	導板	guideway
导边	導緣,前緣	leading edge
导弹护卫舰	飛彈巡防艦	guided missile frigate, FFG
导弹快艇	[導向]飛彈快艇	guided missile boat, FAC (missile)
导弹卫星跟踪测量船	飛彈衛星跟蹤測量船	missile range instrumentation ship
导风罩	導風罩	wind scooper
导管推进器	導罩螺槳	ducted propeller, shrouded propeller
导航	航海	navigation
导航板	導板	guide plate
导航参数	導航參數	navigation parameter
导航杆	導桿	guide rod
导航设备	助航設施	navigational aid
导航声呐	導航聲納	navigation sonar
导航卫星	導航衛星	navigation satellite
导缆孔	[系船]導索管	mooring pipe
导缆器	導索器	fairlead, fairleader
导缆钳	導[纜]索器	chock
导体	導體	conductor

祖 国 大 陆 名	台 湾 地 区 名	英 文 名
导向杆	導桿	guided rod, guide bar
导向滚轮	台式滾輪	pedestal roller
导向索	導索,扶手索	guideline, guide rope
导[向]叶[片]	導葉	guide blade, guide vane
导叶泵	擴散泵	diffuser pump
倒车阀	倒車閥,倒車[保]護閥	astern valve, astern guardian valve
倒车工况		astern condition
倒车喷嘴	倒車噴嘴	astern nozzle
倒车燃气轮机	倒車燃氣渦輪機	astern gas turbine
倒车试验	倒車試航	astern trial
灯船	燈標船	light boat
灯光诱鱼船	誘魚燈船	fish luring light boat
登陆艇	登陸艇	landing craft
登陆战舰艇,两栖作战舰艇	兩栖作戰艦艇	amphibious warfare ships and crafts
登艇灯	[艇筏]乘載照明燈,小艇甲板燈	boat deck light, embarkation lamp, boat deck lamp
登艇梯	登艇梯	boarding ladder
等螺距	定螺距	constant pitch
低磁钢	非磁性鋼	low-magnetic steel, non-magnetic steel
低氢型焊条	低氫焊條	low hydrogen type electrode, hydrogen controlled electrode
低速风洞	低速風洞	low speed wind tunnel
低位通海阀	低位海水吸入閥	low sea suction valve
低温腐蚀	低溫腐蝕	low temperature corrosion
低压空气瓶	低壓空氣瓶	low pressure air bottle
低压透平	低壓渦輪機	low pressure turbine
低压蒸汽发生器	低壓蒸汽產生器	low pressure steam generator
底边舱		bottom side tank
底舱肋骨	艙肋骨	hold frame
底漆	底漆	primer
地球物理勘探船	地球物理探測船	geophysical survey vessel, geophysical survey ship
地质调查船	地質調查船	geological survey vessel, geological survey ship
第二甲板	第二[層]甲板	second deck
点焊	點焊	spot welding
点火	點火	ignition

祖 国 大 陆 名	台 湾 地 区 名	英 文 名
点火器	點火器	igniter, ignition plug
点火提前角	點火提早角	spark advance angle, ignition advance angle
点火延迟	點火遲延	spark lag, ignition lag
点燃式内燃机	引燃機	spark ignition engine
点蚀	斑蝕, 孔蝕	pitting, pitting corrosion
电磁干扰	電磁干擾	electromagnetic interference
电磁计程仪	電磁計程儀	electromagnetic log
电磁兼容性	電磁兼容性	electromagnetic compatibility, EMC
电磁离合器	電磁離合器	electromagnetic clutch
电磁铁	電磁鐵	electromagnet
电磁吸盘	起重磁鐵	lifting magnet
电导率	傳導性, 傳導度	conductivity
电动操舵装置	電動舵機	electric steering gear
电动辅机自动起动装置	電動輔機自動起動裝置	automatic starting installation for electrical motor driven auxiliaries
电动机	電動機, 馬達	motor
电动膜片式气笛	電動氣笛	motor siren
电动势	電動勢	electromotive force, electric motive force
电动液压操舵装置	電動液壓操舵系統	electro-hydraulic steering gear, electro-hydraulic steering system
电动主机传令装置	電動車鍾	electric engine telegraph
电镀	電鍍	electroplating
电工间	電工場	electrician's store, electrical shop
电功率	電功率	electrical power, electric power
电弧点焊	電弧點焊	arc spot welding
电弧电压	電弧電壓	arc voltage
[电]弧焊	電[弧]焊[接]	arc welding
[电]弧焊机	電焊機	arc welding machine, arc welder
电弧钎焊	電弧硬焊	arc brazing
电弧切割	電弧切割	arc cutting
电化学腐蚀	電化學腐蝕	electrochemical corrosion
电化学氧化	陽極處理, 陽極防蝕	anodizing
电极	電極	electrode
电加热	電[加]熱	electric heating
电加热器	電熱器	electric heater
电解质	電解質	electrolyte
电控罗经		electromagnet control gyrocompass

祖国大陆名	台湾地区名	英文名
电缆舱	電纜艙	cable tank
电缆框,电缆筒	穿線環圍,線孔線圍	cable coaming
电缆伸缩箱	電纜箱	cable expansion box, cable box
电缆松紧指示器	電纜鬆緊計	slack meter
电缆筒(=电缆框)		
电力推进	電力推進	electric propulsion
电力推进船	電力推進船	electric propulsion ship
电力推进装置	電力推進設備	electric propulsion plant
电流	電流	current
电偶腐蚀	電蝕	galvanic corrosion
电气试验	電試驗	electric test
电渣焊	電熱熔渣焊接	electro-slag welding
电装	電裝	electric fitting
电子束焊	電子束焊	electron beam welding
电子调速器	電子調速器	electronic governor
电阻焊	電阻焊接	resistance welding
电阻钎焊	電阻硬焊	resistance brazing
垫升风扇	墊升風機	lift fan
吊板	工作吊板,單人吊板	boatswain's chair
吊放回收装置	吊放回收裝置	launch retrieval apparatus
吊放式救生筏		davit launched type liferaft
吊杆	吊桿,起重桿	derrick boom
吊杆安全工作负荷	安全工作負荷,安全使用負荷	safe working load, SWL
吊杆偏角	吊桿旋角	slewing angle
吊杆托架	吊桿托架	derrick rest
吊杆柱	主柱	king post
吊杆装置	吊桿裝置	derrick rig, derrick
吊杆座	鵝頸型吊桿座,吊桿承座	gooseneck bracket, derrick socket
吊缸	吊缸	lift out piston
吊货索	吊貨索	cargo runner, cargo sling
吊货索具	吊貨索具	cargo purchase rigging, cargo wire runner
吊梁	吊梁	lifting beam
吊锚杆	吊錨桿	anchor davit, anchor crane, cat davit, cat head
吊锚索具	吊錨索具,吊錨滑車組	cat-chain, cat tackle
吊艇架	小艇吊架	boat davit

祖 国 大 陆 名	台 湾 地 区 名	英 文 名
吊艇索	小艇吊索	boat fall
吊艇装置	小艇吊放装置	boat handing gear
吊装分析	吊装分析	lifting analysis
碟式油分离机		disc oil separator
顶边舱	[翼]肩舱	top side tank
顶浪	顶浪	head sea
顶索绞车	俯仰绞机(吊杆)	topping winch
[顶]推船	推[型拖]船	pusher, push boat
定常空泡	稳定空泡	steady cavities
定额	定额	norm
定容循环	等容循环	constant volume cycle
定位	定位	locating
定位焊	定位点焊	tack welding
定位桩	柱锚[挖泥船]	spud
定位桩架	定位椿架	spud gantry
定位桩台车	定位椿台车	spud carriage
定压循环	等压循环	constant pressure cycle
动横倾角	动横倾角	dynamical heeling angle
动力定位	动态定位	dynamic positioning
动力系统	动力系统	power system
动平衡	动态平衡	dynamic balancing
动平衡试验	动平衡试验	dynamic balancing test
动倾覆角	动倾覆角	dynamical upsetting angle
动稳定性	动稳度	dynamic stability
动稳性	动稳度	dynamical stability
动稳性曲线	动稳度曲线	curve of dynamical stability
动叶[片]	[转]动叶片	moving blade
动载荷	动态负荷	dynamic load
堵孔	塞孔	plug-hole
渡船	渡船	ferry, ferry boat
端接焊缝	边[缘]焊接	edge weld
端面	边缘	edge
短峰波	短峰波	short crested waves
短路	短路	short circuit
短路电流	短路电流	short-circuit current
断续焊	断续焊法	intermittent welding
锻锤	锻锤	forging hammer
锻接	锻接	forging welding, forge weld

祖 国 大 陆 名	台 湾 地 区 名	英 文 名
锻造比	鍛造比	forging ratio
堆焊	堆焊, 堆焊接	surfacing, build up welding
对称浸水		symmetrical flooding
对动活塞式内燃机	對沖活塞引擎	opposed-piston engine
对接焊	對接焊接	butt welding
对开船体	開體船	split hull
对开耙吸挖泥船	對開式耙吸船	split-type trailing suction dredger, split-trail
对空潜望镜	對空潛望鏡	altiperiscope
对转螺旋桨	對轉螺槳	contrarotating propellers
吨位	噸位	tonnage
吨位丈量	噸位丈量	tonnage measurement
墩	墩	blocks
镦粗	鐓鍛, 鍛粗	upsetting
趸船	躉船	pontoon
钝边	根面(焊接)	root face
多层焊	多層焊接	multi-layer welding
多浮筒系泊系统	多浮筒系泊系統	multi-buoy mooring system
多级预热给水	多級預熱給水	multi-stages feed heating
多甲板船	多甲板船	multi-decked ship
多孔板	多孔板	perforated distribution plate, perforated plate
多能打桩架	多能打樁架	multiple-pile driver tower
多普勒计程仪	都卜勒測程儀	Doppler log
多普勒声呐	都卜勒聲納	Doppler sonar
多体船	多體船	multi-hulled ship, multi-hulled craft
多芯电缆	多芯電纜	multi-core cable
多压式汽轮机	多壓渦輪機	multipressure turbine
多叶舵	多葉舵	multi-bladed rudder
多用途货船	多用途貨船	multipurpose cargo ship, general purpose ship
多支承舵	多舵針舵	multi-pintle rudder
舵	舵	rudder
舵板	舵板	rudder plate
舵臂	舵臂	rudder arm
舵柄	舵柄	tiller
舵柄连杆	舵柄連桿	tiller tie-bar
舵掣	舵韌, 舵制動器	rudder brake

祖国大陆名	台湾地区名	英 文 名
舵承	舵軸承	rudder bearing
舵杆	舵桿	rudder stock
舵杆接头	舵頸接頭	rudder coupling
舵杆扭矩	舵轉矩	rudder torque
舵杆填料函	舵填料函	stuffing box, rudder stuffing box
舵构架	舵肋	rudder frame
舵机	操舵装置	steering gear
舵机舱	舵機艙	steering engine room, steering gear room
舵角	舵角	rudder angle, helm angle
舵角限位器	舵止器,制舵器	rudder stop, rudder stopper
舵角指示器	舵角指示器	electric rudder angle indicator, rudder angle indicator
舵链导轮	導滑車	leading block
舵面积比	舵面積比	rudder area ratio
舵钮	舵針承	rudder gudgeon
舵平衡比	舵平衡面積比	coefficient of balance of rudder, rudder balance area ratio
舵扇	舵柄弧	quadrant
舵设备	舵装置	rudder and steering gear, rudder gear
舵头	上部舵桿,舵頭材	rudder head
舵销	舵針	rudder pintle
舵压力	舵壓	rudder pressure
舵压力中心	舵力中心	center of rudder pressure, center of rudder force
舵叶	舵葉	rudder blade
舵轴	舵軸	rudder axle
舵轴舵	[舵]軸兼[舵]柱型舵	simplex rudder
舵柱	舵柱	rudder post
惰性气体	惰氣	inert gas
惰性气体保护焊	惰氣[金屬]電[弧]焊,金屬惰氣電弧焊	inert-gas arc welding, SIGMA welding
惰性气体发生装置	惰性産生器	inert gas generators, IGG
惰性气体鼓风机	惰氣鼓風機	inert gas blower
惰性气体系统	惰氣系統	inert gas system, IGS

E

祖国大陆名	台湾地区名	英文名
额定功率	额定功率	rated power
额定[输出]功率	额定出力	rated output
二冲程内燃机	二衝程引擎	two-stroke engine
二冲程循环	二衝程循環	two-stroke cycle
二氧化碳灭火器	二氧化碳滅火器	carbon dioxide fire extinguisher

F

祖国大陆名	台湾地区名	英文名
发电机	發電機	generator
发电机电动机系统	列氏電動操作系統	generator-motor system, Ward-Leonard system
发电机组	發電機組	generating set
发电用内燃机	發電機引擎	genset engine, dynamo engine
发火次序	點火順序	firing order
乏汽轮机	排[蒸]汽渦輪機	exhaust steam turbine
阀面研磨	閥面研磨	valve lapping
阀箱	岐管閥	manifold valve
筏	筏	raft
法定检验	法定檢驗	statutory survey
法兰连接	凸緣接合	flange joint
帆船	帆船	sailer, sail boat
帆缆间	帆纜庫,索具庫	hawser store, deck store
帆艇	帆船	sailing boat
反动度	反動度(渦輪機)	degree of reaction
反动式汽轮机	反動式渦輪機	reaction turbine
反射炉	反射爐	reverberatory furnace
反渗透海水淡化装置	逆滲透[海水淡化]裝置	reverse osmosis desalination device, reverse osmosis device
反应堆舱	[核子]反應爐艙	reactor room
反应舵	反動式舵	reaction rudder
方龙骨	條龍骨	bar keel
方艉	平艉	transom stern

祖 国 大 陆 名	台 湾 地 区 名	英 文 名
方艉端面	艉横板,艉横材	transom
方位角	方位角	azimuth angle, azimuth
方向谱	方向频谱	directional spectrum
方向稳定性	方向穩定性	directional stability
方形系数	方塊係數	block coefficient
防冰设施	防冰設備	anti-icing device, anti-icing equipment
防尘	防塵	dust-protected
防护服	防護衣	protective clothing
防护罩	護罩	protective cover
防火舱壁	防火艙壁	fireproof bulkhead, firewall
防火风门	防火擋板	fire damper
防火门	防火門	fire door
防火网	滅焰器	flame arrester
防浪阀	止浪閥	storm valve
防倾肘板	防撓肘板	tripping bracket
防蚀	防蝕	corrosion protection, corrosion prevention
防鼠板	防鼠板	rat guard
防水盖布	油帆布	tarpaulin
防酸性能	耐酸,抗酸	acid resistance, acid proof
防污	防污	anti-fouling
防污涂料	防污漆	anti-fouling paint
防锈涂料	防銹漆,防蝕漆	anti-corrosion paint
防锈油	防銹油	rust preventive oil
防撞舱壁	防碰艙壁	collision bulkhead
放气阀	吹泄閥	blow off valve
放气系统	吹泄系統	blow-off system
放艇安全索	放艇安全索	life rope
飞车	螺槳空車,飛車	propeller racing
飞高	裙底氣隙	hover gap
飞溅	焊濺物	spatter
飞溅区	飛濺區	splash zone
飞溅区腐蚀	飛濺區腐蝕	splash zone corrosion
飞溅阻力	濺水阻力	spray resistance
飞轮	飛輪	fly wheel
飞行甲板	飛行甲板(航艦)	flight deck, flying-off deck
非机动船,非自航船	無動力船	non-powered ship
非金属夹杂	非金屬夾雜物	non-metallic inclusions
非水密舱壁	非水密艙壁	non-watertight bulkhead

祖国大陆名	台湾地区名	英 文 名
非水密门	非水密門	non-watertight door
非自航船（=非机动船）		
废气涡轮增压	渦輪增壓	turbo-charging
分舱吃水	艙區劃分吃水	subdivision draft
分舱因数	艙區劃分因數	factor of subdivision
分舱载重线	艙區劃分載重線	subdivision loadline
分段多层焊	間段焊接法	block sequence welding
分段建造法	分段建造法	sectional method of hull construction
分段舾装		section outfitting
分段装配	［船體］分段裝配	fabrication, section assembly
分断电流	切斷電流	breaking current
分出功率输出装置（=辅助功率输出装置）		
分级卸载	卸載	load shedding
分罗经	子羅經, 羅經複示儀	compass repeater
分配电箱	配電盤, 分電箱	distribution board, distribution box
分体式浮船坞	分段浮塢	sectional dock, sectional floating drydock
分型剂	脫模劑	parting agent
粪便泵	穢水泵, 污水泵	sewage pump
风暴扶手	風暴扶手	storm rails
风道	通風管	air duct
风、浪、流试验水池	風、浪、流試驗水槽	wind wave and current tank
风冷	空氣淬火	air quenching
风雨密门	風雨密門	weather tight door
风雨密性	風雨密性	weathertightness
封闭	封閉	sealing
封舱锁条	鎖緊柄	locking bar
封舱楔	艙口楔	hatch wedge
封舱压条	艙口壓條	hatch batten
封底焊	防漏焊	sealing welding, seal weld
封底焊道	背焊道	back weld, sealing run
峰隙	距波高度	wave clearance
缝隙腐蚀	間隙腐蝕, 隙間腐蝕	crevice corrosion, interstitial corrosion
敷网渔船	敷網漁船	square netter
扶正分析	扶正分析	uprighting analysis
服务航速	航海船速, 營運船速	service speed
浮船坞, 浮坞	浮塢	floating dock
浮力	浮力	buoyancy force, buoyancy

祖国大陆名	台湾地区名	英 文 名
浮力舱	空艙	air tank, void space
浮力曲线	浮力曲線	buoyancy curve
浮锚	海錨	floating anchor, sea anchor
浮式生产储卸油装置	浮式採油储油及卸油	floating production storage offloading, FPSO
浮式生产储油装置	浮式生産储油單元	floating production storage unit, FPSU
浮式生产系统	浮式生産系統	floating production system, FPS
浮态	浮揚狀態	floating condition
浮体	浮體	floating body
浮筒卸扣	浮筒系鈎	buoy shackle, buoy hook
浮筒液位计	浮動液位計	float level gauge
浮坞(=浮船坞)		
浮箱	浮力艙櫃, 浮箱	buoyancy tank, ponton
浮箱式浮船坞		pontoon floating dock
浮心	浮[力中]心	center of buoyancy
浮心曲线	浮[力中]心曲線	curve of centers of buoyancy
浮心纵向坐标	縱向浮[力中]心	longitudinal center of buoyancy
浮性	浮力	buoyancy
浮油回收船	捞油船	oil skimmer
辐流式汽轮机	徑向流渦輪機	radial flow turbine
辅柴油机	輔柴油機	auxiliary diesel engine
辅给水系统	輔給水系統	auxiliary feed system, auxiliary feed line
辅锅炉	輔鍋爐, 副鍋爐	auxiliary boiler, donkey boiler
辅机舱	輔機艙	auxiliary machinery compartment, auxiliary machinery room
辅汽轮机	輔蒸汽渦輪機	auxiliary steam turbine
辅汽轮机组		auxiliary steam turbine set
辅燃气轮机	輔燃氣渦輪機	auxiliary gas turbine
辅助操舵装置	輔[助]操舵裝置	auxiliary steering gear
辅助功率输出装置,分出功率输出装置	輔助功率輸出裝置	power take-off, PTO
辅助舰船,军辅船	輔助艦	auxiliary ship and service craft, auxiliary ship
腐蚀	腐蝕, 銹蝕	corrosion
腐蚀产物	腐蝕生成物	corrosion product
腐蚀疲劳	腐蝕疲勞, 銹蝕疲勞	corrosion fatigue
腐蚀试验	腐蝕試驗	corrosion test
腐蚀裕量	腐蝕裕度	corrosion allowance

祖国大陆名	台湾地区名	英 文 名
负荷	負荷	load
负荷试验	負荷試驗	load test
负载	負載	load
附加质量	附加質量	added mass
附体	附屬物	appendages
附体阻力	附屬物阻力	appendage resistance
附着力	黏著[力],附著[力]	adhesion
复合钢	護面鋼	clad steel
复向期	回復期	reach
复原力臂	扶正力臂	righting lever, restoring lever
复原力矩	扶正力矩,復原力矩, 回復力矩	righting moment, restoring moment
腹板	腹板,大肋骨	web
覆板	加力板,加強板,二重板	doubling plate
覆材甲板	包板甲板,被覆甲板	sheathed deck

G

祖国大陆名	台湾地区名	英 文 名
干船坞	乾[船]塢	dry dock, graving dock
干粉灭火器	乾粉滅火器	powder fire extinguisher
干货船	乾貨船	dry cargo ship
干罗经	乾羅經	dry-card compass, dry compass
干涉	干涉	interference
干湿两用阀装置	乾濕兩用閥裝置	wet and dry pipe valve installation
干舷	乾舷	freeboard
干舷甲板	乾舷甲板	freeboard deck
坩埚炉	坩堝爐	crucible furnace
竿钓渔船	竿釣漁船	pole and line fishing boat
感温式探测器	溫度探測器	thermal detector
感烟式探测器	探煙器	smoke detector
感应电动机	感應馬達,感應電動機	induction motor
感应加热	[電]感應加熱	induction heating
感应钎焊	[電]感應硬焊	induction brazing
感应线圈	探索線圈	search coil
刚性救生筏	硬式救生筏	rigid liferaft
钢船	鋼[殼]船	steel ship
钢筋混凝土船	鋼筋混凝土船	reinforced concrete ship, reinforced con-

祖 国 大 陆 名	台 湾 地 区 名	英 文 名
		crete vessel
钢丝网水泥船	鋼筋水泥船	ferrocement boat，ferrocement vessel，ferrocement ship
港务船	港勤艇	harbour craft
高温腐蚀	高溫腐蝕	high temperature corrosion
高性能船	高性能船	high performance craft
高压[给水]加热器	高壓加熱器	high pressure feed water heater，high pressure heater
高压锅炉	高壓鍋爐	high pressure boiler
高压空气瓶	高壓空氣瓶	high pressure air bottle
高压透平	高壓渦輪機	high pressure turbine
搁墩负荷	擱墩負載	block load
割炬	火焰截割器	cutting torch
格子线	格子	grid
隔板	隔板	division plate
隔舱填料函	艙壁填料函	bulkhead stuffing box
隔离堆焊层	預堆邊焊	buttering
隔膜	隔膜	diaphragm
隔叶块	隔片，間隔件	spacer
各向异性	異向性	anisotropy
给气比	給氣比	delivery ratio
给水	給水，爐水	feed water
给水阀	給水閥	feed water valve，feed valve
给水加热器	給水加熱器	feed water heater
给水软化装置	給水軟化裝置	water-softening plant
根部半径	根[部]半徑	root radius，groove radius
根部焊道	初層焊道	root pass
根部间隙	根隙(焊接)	root gap
根厚	葉根厚度	root thickness
根涡	葉根渦旋	root vortex
工程船	工作船，作業船	working ship，work ship
工地焊接	現場焊接	field welding
工件	工作件	workpiece
工作持续时间	歷時，時段	duration
工作艇	工作船，工作艇	work boat，utility boat
工作行程	工作衝程	working stroke
功率	功率	power
功能	功能	function

祖国大陆名	台湾地区名	英 文 名
拱	拱	arch
供应船	補給船	tender, supply ship
共振	共振	resonance
沟状腐蚀	槽蝕	groovy corrosion, grooving corrosion
构件	構件	member
骨架	骨架,構架	framing
毂径	螺旋毂直徑	hub diameter
毂径比	毂徑比	hub diameter ratio, hub ratio
毂帽	螺槳帽	propeller cap
固定锚	繫留錨,碇泊錨	mooring anchor
固定式内燃机	固定引擎	stationary engine
固化	硬化	curing
固化剂	硬化劑	hardener
固溶热处理	溶液熱處理	solution heat treatment
固体浮力材料	固體浮材	solid buoyancy material
固体压载	固體壓載	solid ballast
固艇索具	艇艄系纜	boat rope
故障	故障	failure
故障电流	故障電流	fault current
故障分析	故障分析	failure analysis
挂车	拖車	trailer
挂舵臂	半懸舵承架	rudder horn
关闭设备	關閉裝置	hull closures, close appliance
观察窗	檢查孔,觀察孔	viewport
管板	管板	tube plate
管结点	管接頭	tubular joint
管卡	管夾,管固定帶	pipe band
管路冲洗		flushing of pipeline
管壳式热交换器	殼管式熱交換器	shell and tube heat exchanger
管形燃烧室	筒形燃燒室	can-type combustor, can-type chamber
管形支柱	圓管柱	tubular pillar
管子吊架	管吊架	pipe hanger
贯通肘板	全通腋板,貫通托架	through bracket
惯性导航系统	慣性導航系統	inertial navigation system
惯性增压	慣性增壓	inertia supercharging
灌水试验	注水試驗	water filling test
光信号	視覺信號	visual signal
龟裂	龜裂	crazing, crack

祖国大陆名	台湾地区名	英 文 名
规则波	規則波	regular wave
辊涂	滾輪式塗漆	roller painting
滚动式舱口盖	滾動[式]艙口蓋	rolling hatchcover
滚翻式舱口盖	單拉[式]艙口蓋	single pull hatchcover
滚卷式舱口盖	捲動[式]艙口蓋	roll stowing hatchcover
滚轮导缆器	滾子導索器	roller fairlead, roller fair-leader
滚装船	滾裝船,駛上駛下船,轆轆船	roll on-roll off ship, Ro/Ro ship, drive on/drive off ship
滚装信道设备	滾裝信道設備	Ro/Ro access equipment
锅炉	鍋爐	boiler
锅炉本体	鍋爐本體	boiler proper, boiler body
锅炉舱	鍋爐艙,鍋爐間	boiler room, fire room
锅炉给水泵	鍋爐給水泵	boiler feed pump
锅炉燃油泵	鍋爐燃油泵	fuel oil burning pump, boiler fuel oil pump
锅炉燃油系统	鍋爐燃油系統	boiler fuel oil system
锅炉水强制循环泵	鍋爐強制循環泵	boiler water forced circulating pump, boiler forced circulating pump
锅炉效率	鍋爐效率	boiler efficiency
锅炉支座	鍋爐鞍座	boiler saddle
锅炉自动控制装置	鍋爐自動控制系統	boiler automatic control system
锅炉座	鍋爐座	boiler foundation, boiler bearer
锅水	[鍋]爐水	boiler water
国际船舶吨位丈量公约	船舶噸位丈量國際公約	International Convention on Tonnage Measurement of Ships
国际船舶载重线公约	國際載重線公約	International Convention on Load Lines, ILLC
国际船舶载重线证书	國際載重線證書	International Load Line Certificate
国际船级社协会	國際船級協會聯合會	International Association of Classification Societies, IACS
国际吨位证书	國際噸位證書	International Tonnage Certificate
国际防止船舶造成污染公约	防止船舶污染國際公約	International Convention for The Prevention of Pollution from ships, MARPOL
国际防止散装运输有毒液体物质污染证书	國際載運散裝有毒液體物質防止污染證書	International Pollution Prevention Certificate for The Carriage of Noxious Liquid Substance
国际防止生活污水污染证书	國際防止污水污染證書	International Sewage Pollution Prevention Certificate

祖国大陆名	台湾地区名	英文名
国际防止油污证书	國際防止油污證書	International Oil Pollution Prevention Certificate, IOPP cert,
国际海上避碰规则	國際海上避碰規則	International Regulations for Preventing Collisions at Sea GSE, COLREGS
国际海上人命安全公约	海上人命安全國際公約	International Convention for The Safety of Life at Sea, SOLAS
国际海事组织	國際海事組織	International Maritime Organization, IMO
国际集装箱安全公约	國際安全貨櫃公約	International Convention for Safety Container, CSC
国际散装运输危险化学品船舶构造与设备规则	國際散裝化學品章程	International Code for the Construction and Equipment of Ships Carrying Dangerous Chemicals in Bulk, LBC code
国际散装运输液化气体船舶构造和设备规则	國際氣體載運船章程	International Code for the Construction and Equipment of Ships Carrying Liquefied Gases in Bulk, IGC code
国际通岸接头	國際岸上接頭	international shore connection
过电流	過[量]電流	over current
过渡舱	過渡艙	transfer chamber, entry locker
过负载	過載,超載	overload
过冷	過[度]冷[卻]	undercooling, supercooling
过热	過熱	overheating, over heat
过热度	過熱度	degree of superheat
过热器	過熱器	superheater

H

祖国大陆名	台湾地区名	英文名
海船	海船,海輪,遠洋船	sea-going ship
海底电缆	海底電纜	submarine cable
海军系数	海軍常數,海軍係數	Admiralty coefficient
海况	海況,海面狀態	sea condition
海流	洋流	current
海上补给	海上整補	replenishment at sea, RAS
海水泵	海水泵	sea water pump
海水淡化装置	海水淡化裝置	sea water desalting plant
海水滤器	海水濾器	sea water filter
海水蒸馏装置	海水蒸餾裝置	sea water distillation plant, sea water distillate plant

祖国大陆名	台湾地区名	英 文 名
海损	海损(保險)	average
海图灯	海圖燈	chart table light, chart lamp
海图室	海圖室	chart room, chart house
海图桌	海圖桌	chart table
海峡[渡]船	海峽船	channel ship, channel steamer
海洋调查船	海洋研究船,海洋調查船	oceanographic research vessel, oceanographic research ship, oceanography survey ship
[海洋]平台	海域平台	platform, offshore unit, offshore platform
海洋污染	海洋污染	marine pollution
氦弧焊	氦[氣電]弧焊接	helium arc welding, heliarc welding
焊层	焊層	layer
焊道	焊珠	bead
焊道下裂纹	焊珠底龜裂	under bead crack
焊点距	焊道間距	weld spacing
焊缝	焊道	weld
焊缝长度	焊道長度	weld length
焊缝金属	焊接金屬	weld metal
焊缝区	焊接金屬面積	weld metal area
焊缝轴线	焊接線	axis of weld
焊根裂纹	根部龜裂	root crack
焊后热处理	焊接後熱處理	postweld heat treatment
焊剂	焊藥,助焊劑	welding flux
焊件	焊件	weldment
焊脚	焊腳(填角焊)	leg
焊接	焊接[法]	welding
焊接操作	焊接操作	welding operation
焊接操作机	自動操控機	manipulator
焊接电弧	焊接電弧	welding arc
焊接电流	焊接電流	welding current
焊接电源	焊接電源	welding power source
焊接工艺参数	焊接條件	welding condition, welding parameter
焊接技术	焊接技術	welding technique
焊接夹具	夾具	fixture
[焊接]接头	焊接接頭	welding joint
焊接顺序	焊接順序	welding sequence
焊接速度	焊接速度	welding speed
焊接通电时间	焊接時間	weld time

祖国大陆名	台湾地区名	英文名
焊接性	可焊性	weldability
焊接应力	焊接應力	welding stress
焊矩	吹把	torch
焊区腐蚀	焊道[晶間]腐蝕	weld decay, weld corrosion
焊条	被覆焊條	covered electrode
焊条夹持端	焊條裸端	bare terminal, exposed core
焊条压涂机	焊條塗藥機	welding rod extrusion press
焊趾	焊接縫突趾	toe of weld
焊趾裂纹	趾裂痕(焊接)	toe crack
航标船	浮標勤務船	buoy tender
航标起重机	燈浮標起重機	light buoy crane
航标艇	設標艇	dan boat
航次	航次	voyage
航道测量船	測量船	hydrographic survey vessel, hydrographic survey ship, survey ship
航空母舰	航空母艦	aircraft carrier, aeroplane carrier
航速	航速	ship speed
航速试验	速率試車	speed trial
航天测量船		space tracking ship
航线	航線	shipping route, trade route
航向	航向	course
航向保持性	航向保持	course keeping quality, course keeping
航行补给船		underway replenishment ship
航行灯	航行燈	running light, navigation light
航行灯控制器	航行燈指示器	navigation light indicator, running light indicator
航行试验	[海上]試航	sea trial
航行信号设备	航行信號設備	navigation signal equipment
号灯	號燈	ship's light
号笛控制装置	號笛控制器	whistle and siren control system, whistle controller
号旗	信號旗	signal flag, code flag
号钟	號鍾	bell
核动力船	核能動力船	nuclear [-powered] ship
核动力装置	核子動力設備	nuclear power plant
核潜艇	核子潛艇	nuclear submarine
黑玻璃	濾光玻璃	filter glass
黑水	污水	black water

祖 国 大 陆 名	台 湾 地 区 名	英 文 名
黑心可锻铸铁	黑心展性鑄鐵	black heart malleable cast iron
桁材	桁,縱梁	girder
桁架桅	籠形桅	cage mast
桁拖渔船	側拖網漁船	beam trawler
横波	橫波	transverse wave
横舱壁	橫向艙壁	transverse bulkhead
横荡	橫移	swaying, sway
横舵柄	橫舵柄	rudder yoke, yoke
横骨架式	橫肋系統	transverse framing system, transverse frame system
横焊	橫[向]焊[接]	horizontal position welding, horizontal welding
横距	回旋橫距	transfer
横浪	橫浪,舷浪	beam sea
横漂	漂流,漂移	drift
横剖面	[橫]剖面	transverse sections
横剖线		body lines
横倾	傾斜,偏斜,傾側	list, heel
横倾角	橫傾角	angle of list, angle of heel
横倾力矩	傾側力矩	heeling moment
横甩	橫甩,橫轉	broaching
横稳心	橫定傾中心	transverse metacenter
横稳性	橫向穩度	transverse stability
横向滑道	橫向滑道	side slipway
横向强度	橫向強度	transverse strength
横向强制摇荡装置	橫向強制搖擺裝置	transverse forced oscillation device
横向下水	橫向下水	side launching
横摇	[船身]橫搖	rolling
横摇角	橫搖角	roll-angle, rolling angle
横摇扭矩	橫搖力矩	rolling torsional moment, rolling moment
横摇阻尼	橫搖阻尼	rolling damping
横移	側滑	sideslip
横张索	吊架跨索	davit span
横重稳距	橫向定傾高	transverse metacentric height
红星火箭	火箭信號	rocket star signal, rocket signal
喉深	喉深(焊接)	throat depth
后备断路器	備用斷路器	back-up breaker
后冷却器	後冷卻器	aftercooler

祖国大陆名	台湾地区名	英文名
后热	後熱（焊接）	postheat, post heating
后体	後半段船體	after body
后桅	後桅	after mast
呼吸阀	呼吸閥	breather valve
互换性	互換性	interchangeability
护卫舰	巡防艦	frigate
护卫艇	護航艇	escort boat
护舷材	護舷材	fender
滑道	下水滑道,下水台	launching way
滑道末端压力	軌端壓力（下水）	way end pressure
滑道摇架	滑道托架	slipway turn cradle, slipway cradle
滑道转盘	滑道轉盤	slipway turntable
滑块	滑塊	slide, sliding block
滑行艇	滑行[快]艇,水上快艇,滑航艇	planing boat, glider
滑油泵	[潤]滑油泵	lubricating oil pump
滑油舱	[潤]滑油櫃	lubricating oil tank
滑油加热器	滑油加熱器	lubricating oil heater
滑油间歇净化	滑油間歇淨化	lubricating oil batch purification
滑油净化系统	滑油淨化系統	lubricating oil purifying system
滑油冷却器	[潤]滑油冷卻器	lubricating oil cooler
滑油滤器	[潤]滑油過濾器	lubricating oil filter
滑油输送泵	滑油輸送泵	lubricating oil transfer pump
滑油系统	滑油系統	lubricating oil system
滑油消耗率	滑油消耗率,單位耗滑油量	specific lubricating oil consumption
滑油泄放柜	滑油泄放櫃	lubricating oil drain tank
滑油泄放系统	滑油泄放系統	lubricating oil drain system
化学反应式灭火器	化學滅火器	chemical reaction fire extinguisher, chemical fire extinguisher
化学品船	化學液體船,化學品船	chemical tanker, chemical carrier, chemical cargo ship
化学清洗	化學清洗	chemical cleaning
划桨救生艇		oar-propelled lifeboat
划线	劃線	laying out, layout
环壁阻力损失	環周損失	annulus drag loss
环境监测船		environmental monitoring vessel, environmental monitoring ship

祖 国 大 陆 名	台 湾 地 区 名	英 文 名
环境温度	環境溫度,周圍溫度	ambient temperature
环境压力	環境壓力,周圍壓力	ambient pressure
环境载荷	環境負荷	environmental load
环照灯	環照燈	all-round light
缓冲柜	緩衝櫃	surge tank
缓冲器	緩衝器	buffer
换能器	換能器	transducer
换热面积	受熱面積	heating surface area
换水孔	換水孔	exchanging water hole
换算箱	二十英尺貨櫃當量	twenty-feet equivalent units, TEU
换向机构	逆轉裝置	reversing gear
换向时间	換向時間	reversing time
换向试验	逆轉試驗	reversing test
晃荡	沖激	sloshing
灰分	含灰量	ash content
灰铸铁	灰[口]生鐵	gray cast iron, gray pig iron
回火	回火	back fire, flashback
回波	回波(電)	echo
回程	回行衝程	return stroke
回火色	回火色	temper color
回流扫力	環流驅氣,環流掃氣	loop scavenging
回热器	再生器	regenerator
回热式汽轮机	再生式渦輪機	regenerative turbine
回声测深仪	回聲測深儀	echo sounding, echo sounder, echo sounding machine
回转横倾角	回轉傾側	heel on turning
回转迹线	回旋跡線	turning path
回转绞车	[吊桿]回旋絞車	slewing winch
回转区域	回旋圈	turning circle
回转区域半径	回旋圈半徑	turning circle radius
回转式空气预热器	再生式空氣加熱器	rotary air heater, regenerative air heater
回转试验	回旋試驗	turning test, turning trial
回转直径	回旋直徑	steady turning diameter, tactical diameter
回转中心	回轉中心	center of turning circle
回转周期	回旋周期	turning period
汇流排	匯流排	busbar
混合器	混合器,調合機	mixer
混合式凝汽器	混合冷凝器	mixing condenser

祖 国 大 陆 名	台 湾 地 区 名	英 文 名
混合式水加热器	混合式水汽加熱器	vapor and water mixing heater
混合室	混合室	mixing chamber
混合循环	混合循環	mixed cycle
混凝土搅拌船		floating concrete mixer
混凝土重力式平台	混凝土重力式鑽油台	concrete gravity platform
活动梯步	舷梯活動踏步	feathering tread, feathering step
活塞	活塞	piston
活塞平均速度	活塞平均速度	mean piston speed
活鱼运输船	活魚運輸船,活魚船	live fish carrier
火箭发射装置	火箭發射器	rocket launcher
火箭降落伞火焰信号	火箭式降落傘照明彈	rocket parachute flare signal, rocket parachute flare
火炮	火炮	artillery
火箱	火箱	fire box
火星熄灭器	火花防止器	spark arrester
火焰除锈	火焰清除	flame cleaning
火焰淬火	火焰硬化	flame hardening
火焰管	火管	flame tube
火焰监测器	火焰探測器	flame detector
火焰气刨	火焰開槽	flame gouging
火灾报警探测器	火警探測器	detector for fire alarm system, fire detector
火灾警报器	火警警報器	fire alarm sounder, fire alarm
货舱	貨艙	cargo hold, cargo space
货舱工作灯	[裝卸]貨[照明]燈	cargo light, cargo lamp
货舱口	貨艙口	cargo hatch
货舱容积	貨艙容量	cargo capacity, hold capacity
货船	貨船,貨輪	cargo ship, cargo vessel, freighter, cargo carrier, cargo boat
货船安全证书	貨船安全證書	Cargo Ship Safety Certificate
货船构造安全证书	貨船安全構造證書	Cargo Ship Safety Construction Certificate
货船设备安全证书	貨船安全設備證書	Cargo Ship Safety Equipment Certificate
货船无线电安全证书	貨船安全無線電證書	Cargo Ship Safety Radio Certificate
货棚	風雨棚	shed
货油泵	貨油泵	cargo oil pump
货油舱	貨油艙	cargo oil tank
货油舱加热系统	貨油加熱系統	cargo oil tank heating system, cargo oil heating system
货油舱扫舱系统	貨油艙收艙系統	cargo oil tank stripping system

祖 国 大 陆 名	台 湾 地 区 名	英 文 名
货油舱透气系统	貨油艙通氣系統	cargo oil tank venting piping system, cargo oil tank venting system
货油舱洗舱系统	貨油艙洗艙裝置	cargo oil tank cleaning system, cargo oil tank cleaning installation
货油阀	貨油閥	cargo oil valve
货油软管	貨油軟管	cargo oil hose
货油装卸系统	貨油裝卸系統	cargo oil handing system, cargo oil pumping system
货油装卸总管	貨油總管	cargo oil main line, cargo oil transfer main pipe line

J

祖 国 大 陆 名	台 湾 地 区 名	英 文 名
机舱	機艙	engine room
机舱布置	機艙佈置	engine room arrangement
机舱辅机		engine-room auxiliary machinery
机舱集控室	機艙控制室	engine control room
机舱集控台		centralized control console of engine room
机舱棚	機艙棚,機艙天罩	engine room casing
机舱涂料	機艙塗料	engine compartment paint, engine compartment coating
机舱自动化	機艙自動化	engine room automation
机动船,自航船	機動船,動力船,自航船	power-driven ship, self-propelled vessel, power-driven vessel
机动救生艇	動力救生艇,馬達救生艇	motor lifeboat
机帆船	機帆船	power-sail ship, motor sailer
机库	[飛]機庫	hangar
机体	機體,引擎體	engine block
机械除锈	機械除銹	mechanical rust removal
机械负荷	機械負荷	mechanical load
机械加工	機械加工	machining
机械式调速器	機械[式]調速機	mechanical governor
机械通风	機械通風	mechanical draft, mechanical ventilation
机械效率	機械效率	mechanical efficiency
机械噪声	機械噪音	mechanical noise
机修间	工場	work shop

祖 国 大 陆 名	台 湾 地 区 名	英 文 名
机柱	柱	column
机装	輪機舾裝	machinery fitting
机座	機座	bed plate, engine bed, engine seat
积木式电缆填料盒	多管穿線板	multi-cables transit, MCT
积载	裝載	stowage
积载因数	積載因數	stowage factor
基地勤务船	港勤艇	service craft, service boat
基面	基準面	base plane, datum level
基频	基[本]頻[率],最低頻率,一階頻率	fundamental frequency
基线	基線,基準線	base line, BL datum line
基准分段		basic section
激光测速仪	雷射測速儀	Laser velocimeter
激流装置	激紊裝置	turbulence stimulator
激振机	激振機	vibration exciter
激振试验	激振試驗	exciter test
级效率	級效率	stage efficiency
极地考察船	極地考察船	polar expedition ship
极化	極化	polarization
极区船		arctic vessel
集电环	滑環	slip ring
集中控制	集中控制	centralized control
集装箱船	貨櫃船	container ship
集装箱舱	貨櫃艙	container hold
集装箱吊具	貨櫃吊具	spreader, container spreader, container lifting spreader
集装箱积载图	貨物裝載圖	stowage plan
几何相似船模	幾何相似模型	geometrically similar ship models, geometrically similar model
[几何]压缩比	壓縮比	compression ratio
计程仪	計程儀	log
计量器具	量測儀器	measuring instruments
计算机辅助工艺规程编制	電腦輔助制程規劃	computer-aided process planning, CAPP
计算机辅助设计	電腦輔助設計	computer-aided design, CAD
计算机辅助制造	電腦輔助製造	computer aided manufacturing, CAM
计算机集成制造系统	電腦整合製造系統	computer integrated manufacturing system, CIMS

祖 国 大 陆 名	台 湾 地 区 名	英 文 名
计算载荷	設計負荷	design load
继电器	繼電器,替續器	relay
加大链环	加大鏈環	enlarged link
加工拖网渔船	拖網加工船	factory trawler
加工余量	機制裕度	machining allowance, machine finish allowance
加强层	加強材	reinforcement
加强筋	加強材,防撓材	stiffener
加速腐蚀试验	加速腐蝕試驗	accelerated corrosion test
加速燃气轮机	助力燃氣渦輪機	booster gas turbine
夹具	夾具,緊固具	jigs and fixtures, gripgear
夹杂物	夾雜物(焊接)	inclusion
甲板	甲板	deck
甲板边板	甲板緣[厚]板	deck stringer
甲板窗	甲板透光玻璃	deck light
甲板防滑涂料	甲板防滑塗料	anti-skid deck paint, anti-slip deck paint
甲板敷料	甲板被覆	deck covering
甲板高度		altitude above deck
甲板横桁	甲板深橫桁	deck transverse
甲板加压舱	船上加壓室(潛水作業)	deck compression chamber
甲板间舱	甲板間艙	tweendeck space, tweendeck cargo space
甲板间肋骨	甲板間肋骨	tweendeck frame
甲板排水口	甲板排水孔	deck scupper
甲板洒水系统	甲板灑水系統	deck sprinkle system
甲板室	甲板室	deck house, castle
甲板水封	甲板水封	deck water seal
甲板梯	甲板梯	deck ladder
甲板通岸接头	通岸接頭	shore connection
甲板涂料	甲板漆	deck paint
甲板线	甲板線(載重線標誌)	deck line
甲板淹湿	甲板上浪,甲板濺濕	deck wetness
甲板中线	甲板中線	deck center line
甲板纵骨	甲板縱材	deck longitudinal, deck runner
甲板纵桁	甲板縱梁	deck girder
驾驶甲板	航海甲板,橋樓甲板	navigation deck, bridge deck
驾驶室	駕駛室,[操]舵房	wheel house, navigation bridge
假舱壁		assembly frame

祖 国 大 陆 名	台 湾 地 区 名	英 文 名
假底	[水槽]活動底	false bottom
假毂	假轂	dummy propeller boss
假模	假模	dummy model
尖舭	銳稜舭[線],硬稜舭	hard chine
尖舱肋骨	尖艙肋骨	peak frame
肩	水線肩部	shoulder
减机运转试验	停機試驗	engine cut off test
减轻孔	減輕孔	lightening hole
减速齿轮箱	減速裝置	reduction gear
减温器	減溫器,調溫器	attemperator
减摇装置控制设备	穩度器控制裝置	stabilizer control unit
减轴运转试验	減軸運轉試驗	shaft cut off test
剪板机	截斷機	plane shear, guillotine shear
检疫船	檢疫船	quarantine vessel
碱脆	鹼性脆化	caustic embrittlement
碱度	鹼度	alkalinity
[碱]煮炉	煮煉(除去爐中油脂雜質)	boiling out
舰艇,军舰	軍船	naval ship, warship
桨毂	[螺槳]轂	Hub, boss
桨毂空化	轂空化	hub cavitation
桨叶	[槳]葉	blade
交错断续角焊缝	交錯間斷式填角焊	staggered intermittent fillet weld
交流电动机	交流電動機	alternating current motor
交流发电机	交流發電機	alternating current generator
胶衬泥泵	膠襯泥泵	rubberized dredge pump
角焊	填角焊[接]	fillet welding
角焊缝	填角焊道	fillet weld
绞车	絞車,絞機	winch
绞缆绞车	卷索絞機	warping winch
绞缆筒	卷索鼓	warping end
绞盘	絞盤	capstan
矫直机	矯直器	straightening machine, straightener
铰接塔(=铰接柱)		
铰接柱,铰接塔	關節連接柱	articulated column
铰吸式挖石船	鉸刀吸入式挖泥船	rock cutter suction dredger, cutter suction dredger
校中	對中線	centering

祖 国 大 陆 名	台 湾 地 区 名	英 文 名
校准	校準	calibration
接触器	接觸器	contactor
接地	接地	earthing, grounding, earth, earth connection
接地电路	接地回路	earthed circuit, ground return circuit
接地电阻	接地電阻	resistance of an earthed conductor, ground resistance
接地极	接地棒	earth electrode, ground rod, earth bar
接口(＝界面)		
接收换能器(＝水听器)		
节点	節點	node
节流调节	節流調速	throttle governing
结构吃水	强度[計算]吃水	scantling draft
结构噪声	結構噪音	structure borne noise
结合力	黏著[力],附著[力]	adhesion
结胶	膠結	gumming
介质强度试验	介質強度試驗	dielectric test, dielectric strength test
界面,接口	界面,接口	interface
金属覆盖层	金屬覆蓋層	metal coating
金属扣合	金屬扣合	tie insert
金属喷镀	金屬噴敷	metal spraying
金属热浸镀	金屬熱浸鍍	metal hot dipping
襟翼	襟翼	flap
襟翼舵	襟翼舵	flap-type rudder, flap rudder
进给量	供給	feed
进流段	艏入水段	entrance
进流段长	艏入水段長	length of entrance
进气设备(＝进气装置)		
进气温度	進氣溫度	intake temperature, intake-air temperature
进气行程	吸氣衝程,吸入衝程	suction stroke
进气压力	進氣壓力	intake pressure
进气装置,进气设备	進氣裝置	air inlet unit
进水角	浸水角,泛水角	flooding angle
进速比	前進比(螺槳)	advance ratio
进速系数	前進係數	advance coefficient
进坞	進塢	docking
进相机	進相機(電)	phase ＝advancer
浸沾钎焊	熱浸硬焊	dip brazing

祖 国 大 陆 名	台 湾 地 区 名	英 文 名
经济功率	經濟功率	economical power
经济转速	經濟船速	economical speed
晶间腐蚀	粒間腐蝕	intergranular corrosion
晶粒度	粒度	grain size
精度管理	精度控制	accuracy control
井口平台	井口平台	wellhead platform
净吨位	淨噸位	net tonnage
净功率	淨馬力	net power, net horsepower
净化器	淨化器	purifier
静平衡试验	靜平衡試驗	static balance test
静水力曲线	靜水[性能]曲線圖	hydrostatic curve
静水弯矩	靜水彎矩	still water bending moment
静水压力释放器	釋放裝置	hydrostatic release unit, release device
静稳性	靜穩度	statical stability
静稳性曲线	靜穩度曲線	statical stability curve, curve of static stability
静叶[片]	定子輪葉	stationary blade, stator blade
救捞驳	救難艇	salvage barge, salvage boat
救生登乘梯	[艇筏]承載梯	embarkation ladder
救生筏	救生筏	liferaft
救生服	浸水衣	immersion suit
救生浮	[兩面用硬式]救生筏	life float
救生圈	救生圈	lifebuoy
救生圈自发烟雾信号	救生圈自發煙霧信號	lifebuoy self-activating smoke signal
救生设备	救生設備	life-saving appliance
救生属具	救生屬具	life-saving appliance
救生索	救生索	life-line
救生艇	救生艇	lifeboat
救生艇乘员定额	救生艇容載量	carrying capacity of lifeboat
救生衣	救生衣	lifejacket
救生衣灯	救生衣燈	lifejacket light
救生衣柜	救生衣櫃	lifejacket chest
救生钟	救生鍾	rescue bell
救助泵	救難泵	salvage pump
救助船	救難船	rescue ship
救助艇	救難艇	rescue boat
就地控制	現場控制	local control
巨型油轮	極大型原油輪,巨型油	very large crude oil carrier, VLCC

祖 国 大 陆 名	台 湾 地 区 名	英 文 名
	輪	
拘束船模试验	拘束模型試驗	captive model test
居住舱室	房艙,艙室	cabin
局部舱壁	部分艙壁	partial bulkhead
局部腐蚀	局部腐蝕	localized corrosion
局部强度	局部強度	local strength
局部振动	局部振動	local vibration
矩形窗		rectangular window
卷纲机	繩索捲盤,捲索盤	rope reel
卷筒	捲索鼓,鼓輪	drum, hoisting drum
卷筒铺管驳	捲盤駁船(海洋工程)	reel barge
卷网机	捲網機	net drum
绝对计程仪	絕對計程儀	speed over the ground log
绝缘材料	絕緣材料	insulating material
绝缘层	絕緣	insulation
绝缘电缆	絕緣電纜	insulated cable
绝缘电阻	絕緣電阻	insulation resistance
军辅船(=辅助舰船)		
军舰(=舰艇)		

K

祖 国 大 陆 名	台 湾 地 区 名	英 文 名
开敞式救生艇	敞式救生艇	open lifeboat
开关	開關	switch
开式给水系统	開式給水系統	open feed system
抗滑稳定性	抗滑穩定性	stability against sliding
抗滑桩	抗滑樁	spud for antislip
抗拉强度	拉伸強度	tensile strength
抗谐鸣边	抗諧鳴邊(螺槳)	anti-singing edge
可变载荷	變動負荷,變動負載	variable load
可达性	可達性	accessibility
可锻铸铁	展性鑄鐵	malleable cast iron
可浸长度	可浸長度	floodable length
可控螺距螺旋桨	[可]控[螺]距螺槳, [可]變螺距螺槳	controllable-pitch propeller
可弃压载	可棄壓載	droppable ballast
可调压载水舱	可變壓載艙	variable ballast tank

祖 国 大 陆 名	台 湾 地 区 名	英 文 名
可移式泵	移動式泵	portable pump
可用性	可用性	availability
客舱	客艙	passenger cabin
客船	客船	passenger ship, passenger vessel
客船安全证书	客船安全證書	Passenger Ship Safety Certificate
客货船	客貨輪	passenger-cargo ship
空船重量	輕載	light weight
空船重量分布	空船重量分佈	light weight distribution
空负荷试验	空載試驗,無負載試驗	no-load test
空负荷运行	無負荷運轉	no-load operation, no-load running
空隔舱	堰艙	cofferdam
空化	空化	cavitation
空化螺旋桨	空化螺槳	cavitating propeller
空化数	空化數	cavitation number
空冷	氣冷	air cooling
空泡	氣泡	cavity
空气抽除装置	抽氣器	air exhauster, air extractor
空气锤	[空]氣錘	pneumatic hammer, air hammer
空气管	空氣管	air pipe
空气管头	空氣管頭	air pipe head
空气加热器	空氣加熱器	air heater, air preheater
空气冷却器	空氣冷卻器	air cooler
空气滤器	空氣過濾器,空氣濾清器	air filter
空气螺旋桨	空氣螺槳	air screw
空气箱	水密空氣箱	watertight aircase
空气消耗率	空氣消耗率,單位耗氣量	specific air consumption
空气预热器	空氣預熱器	air heater, air preheater
空气阻力	空氣阻力	air resistance
空蚀	空[泡沖]蝕	cavitation erosion
空心叶片	空心葉片	hollow blade
空心轴	空心軸	hollow shaft
空载吃水	輕載吃水	light draft
空载电压	無負載電壓	no-load voltage
[空载]电压降低装置	減壓裝置	voltage reducing device
空载运行	無負荷運轉	no-load operation, no-load running
控制继电器	控制繼電器,控制電驛	control relay

祖 国 大 陆 名	台 湾 地 区 名	英 文 名
控制器	控制器	controller
控制系统	控制系统	control system
控制站	控制站,控制室	control station
扣板	角牵板	gusset plate
跨距	跨距	span
宽深比	寬深比	breadth depth ratio
矿砂船	礦砂船	ore carrier
矿油船	礦砂與油兼用船	ore/oil carrier
扩散口	擴散器	diffuser
扩压器	擴散器	diffuser

L

祖 国 大 陆 名	台 湾 地 区 名	英 文 名
垃圾船	垃圾艇	garbage boat
垃圾清扫船		garbage cleaning vessel
垃圾运送槽	垃圾斜槽	garbage chute, garbage shoot
拉杆	拉桿	drag link
拉筋	系索	lacing wire
拉索塔	張索塔(鑽油台)	guyed tower
拉削	拉削	broaching
栏杆	欄桿	railing, rail
蓝脆	藍脆性	blue shortness
缆索	大索	hawser
缆索卷车	纜索捲車	rope storage reel
廊道	廊,通道	gallery
浪向	浪向	sea direction
捞雷船	捞雷船	torpedo recovery ship
老化	老化	ageing
雷达截面	雷達截面積	radar cross section, RCS
雷达桅	雷達桅	radar mast
雷达信标	雷[達示]標	radar beacon
雷康	雷[達示]標	racon
雷锚	雷錨	mine anchor
肋板	底肋板	floor
肋骨	肋骨	frame
肋骨框架	肋骨圈	frame ring
肋距	肋骨間距	frame spacing, frame space

祖国大陆名	台湾地区名	英文名
A类机器处所	甲種機艙空間	machinery space of category A
棱形系数	棱形係數,[縱向]棱塊係數	prismatic coefficient
冷藏船	冷藏船,冷凍船	refrigerator ship, refrigerated carrier, cold storage boat
冷藏货舱	冷藏貨艙	refrigerated cargo hold
冷藏库	冷藏庫,冷凍庫	refrigerating chamber, cold store
冷处理剂	冷凍劑,冷媒	refrigerant
冷脆	冷脆性	cold brittleness
冷裂	冷破裂	cold cracking
冷却倍率	冷卻[速]率	cooling rate
冷却水泵	冷卻水泵	cooling water pump
冷却[水]管	冷卻管,冷凝管	condenser tube, cooling pipe
冷却叶片	冷卻葉片	cooled blade, cooling blade
冷态起动	常溫起動,冷車起動	cold starting
冷油器	油冷[卻]器	oil cooler
冷作	冷作	cold work
离心泵	離心泵	centrifugal pump
离心浇注	離心鑄造法	centrifugal pressure casting, centrifugal casting
离心离合器	離心離合器	centrifugal clutch
离心[式]压气机	離心壓縮機	centrifugal compressor
理网机	理網機	net shifter
理想循环	理想循環	ideal cycle
立管	升導管	riser
立式锅炉	立式鍋爐	vertical boiler
立式内燃机	立式引擎	vertical engine
立体分段		three-dimensional unit
立柱	柱	column
励磁变阻器	磁場變阻器	field rheostat
励磁机	勵磁機,激磁機	exciter
利用船体作负极回路的直流单线系统	直流單線船體負極回路系統	negative hull-return D. C. single-wire system, negative hull-return D. C. single system
连杆	連桿	connecting rod
连接链环	連接鏈環	lugless joining shackle
连接桥	連橋	cross structure
连接卸扣	連接環	joining shackle

祖国大陆名	台湾地区名	英 文 名
连续波	連續波	continuous wave
连续电流	連續電流	permanent current, continuous current
连续构件	連續構件	continuous member
连续焊缝	連續焊接	continuous weld
连续铸造	連續鑄造	continuous casting
联合动力装置	複合式動力裝置	combined power plant
联锁机构	連鎖裝置	interlocking device, interlocking gear, in-terlock arrangement
链斗挖泥船	戽斗挖泥船	bucket dredger
链节	錨鏈長度,錨鏈一節	length of chain cable, shackle
链式输送机	鏈條輸送機	chain conveyor
梁拱	拱高	camber
梁肘板	梁肘板,梁腋板	beam knee
粮食库	糧食庫,給養艙	provision store
两段造船法	兩船段建造法	two-part hull construction
两栖作战舰艇(＝登陆战舰艇)		
亮度	亮度	luminance
了望台	桅桿[了]望台	crow's nest
列板	列板,板列	strake
猎雷	獵雷	mine hunting
猎雷舰艇	獵雷艦,獵雷艇	minehunter
猎潜艇	驅潛艇	submarine chaser
裂纹敏感性	裂縫敏感度	crack sensitivity
裂纹试验	破裂試驗	cracking test
临界舵角	臨界舵角	stalling rudder angle
临界角	臨界角	critical angle, danger angle
临界空化数	臨界空化數	critical cavitation number
临界转速	臨界速率,共振轉速	critical speed, resonance speed
淋水试验	淋水試驗	water pouring test
灵敏度	靈敏度,敏感度	sensitivity
溜泥槽	滑槽	chute
流[刺]网渔船	流網漁船	drift netter, drifter, drift fishing boat
流动显示	流場觀察,流場可視化	flow visualization
流化床锅炉	流動床[燃燒]鍋爐	fluidized bed boiler, fluidized bed firing boiler
流量计	流量計	flow meter
流量系数	流量係數	flow coefficient

祖国大陆名	台湾地区名	英文名
流水孔	排泄孔	drain hole
流速计	速度計,測速器	velocity meter, velocimeter
流线型舵	流線型舵	streamline rudder, hydrofoil rudder
流型	流型	flow pattern
六分仪	六分儀	sextant
龙骨	龍骨	keel
龙骨墩	中心墩	keel block
龙骨线	龍骨線	keel line
龙骨翼板	龍骨翼板列,A 板列	garboard strake
漏汽损失	泄漏損失	leakage loss
漏水口	排水孔	scupper
露点	露點	dew point
炉舱棚	鍋爐艙棚,鍋爐艙圍壁	boiler room casing
炉膛	爐膛	furnace
炉膛容积	爐內容積	furnace volume
炉中钎焊	爐內硬焊	furnace brazing
旅游船,游览船	遊覽船	excursion vessel, excursion boat
铝青铜	鋁青銅	aluminum bronze
履带式布缆机	履帶式布纜機	liner cable laying machine
滤器	過濾器	filter
滤油器	淨油機	oil purifier
滤油设备	濾油設備	oil filtering equipment
轮盘	圓盤	disc
罗方向	羅經方位[角]	compass bearing
罗航向	羅經航向	compass course
罗经,罗盘	羅經	compass
罗经甲板	羅經甲板	compass deck
罗兰 A	羅遠 A 導航系統	Loran A
罗兰 C	羅遠 C 導航系統	Loran C
罗兰－惯性导航系统	羅遠－慣性導航系統	Loran inertial navigation system
罗盘(＝罗经)		
螺距	螺距	pitch
螺距比	節圓直徑比	pitch ratio
螺距角	螺距角	pitch angle
螺纹接头连接	活管套接頭	union joint
螺线试验	蝸旋試驗	spiral test
螺旋风管	螺旋風管,螺旋導管	spiral duct
螺旋桨	螺[旋]槳	screw propeller

祖 国 大 陆 名	台 湾 地 区 名	英 文 名
螺旋桨敞水效率	螺槳單獨效率	open water propeller efficiency
螺旋桨船	螺槳船	screw ship, propeller vessel
螺旋桨平面	螺槳平面	propeller plan
螺旋桨推入力	安裝螺槳壓擠力	propeller fitting force
螺旋桨推入量	安裝螺槳壓擠量	propeller pull-up distance
螺旋桨尾流	螺槳尾流，［螺槳］艉流	propeller race, slipstream
螺旋桨轴,艉轴	艉［管］軸	screw shaft, propeller shaft, tube shaft, stern shaft
螺旋桨锥孔研配	螺轂孔研配	scraping of propeller boss
螺旋桨纵倾角	後傾角(螺槳)	rake angle of propeller, rake angle
螺旋扣	緊索螺釘	rigging screw
螺旋输送机	螺旋輸送機	screw conveyor
螺旋压力机	螺［旋］壓機	screw press
螺柱焊	柱焊,嵌柱焊接	stud welding
裸船体	裸船殼	bare hull, naked hull
裸置管道	裸置管路	unburied pipeline

M

祖 国 大 陆 名	台 湾 地 区 名	英 文 名
码头	碼頭	wharf
埋弧焊	潛弧焊,潛溶焊	submerged arc welding, submerged melt arc welding
埋设深度	埋設深度	laying depth
埋艏	艏潛,埋艏	plough-in
脉冲重复频率	脈衝複現頻率	pulse repetition frequency
脉动空泡	脈動空泡	pulsating cavity
满舵舵角	滿舵角	hard-over angle
满载吃水	滿載吃水	loaded draft, load draught
满载排水量	滿載排水量	full load displacement
满载水线	載重水線	load waterline
满载水线长		load waterline length
盲板法兰	管口蓋板	blind flange
盲孔	盲孔	non-penetrated hole, blind hole
盲区	盲區	dead zone, blind zone
毛刺	毛口	burr
毛坯	毛坯	blank
锚	錨	anchor

祖 国 大 陆 名	台 湾 地 区 名	英 文 名
锚泊定位		anchor moored positioning
锚泊设备		anchoring equipment
锚泊状态	錨泊	anchored condition, anchoring
锚床	錨床	anchor bed
锚灯	錨泊燈	anchor light
锚地	錨[泊]池	anchorage
锚端链节	轉環	swivel piece
锚浮标	錨位浮標	anchor buoy
锚干	錨幹	anchor shank, anchor shaft
锚横杆	錨桿	anchor stock
锚架		anchor rack
锚具	錨具	ground tackle
锚缆		anchor hawser
锚链	錨鏈	anchor chain
锚链舱	錨鏈艙	chain locker
锚链管	錨鏈管,錨鏈筒	chain pipe
锚链轮	鏈輪	cable lifter, wildcat
锚链筒	錨鏈筒	hawse pipe
锚链转环	轉環	swivel
锚球	錨球	anchor ball
锚索	錨索,錨纜	anchor cable
锚卸扣	錨接環	anchor shackle
锚穴	嵌錨穴	anchor recess
锚抓力	錨抓著力	anchor holding power
锚爪	錨爪	anchor fluke
锚爪袭角	攻角	angle of attack
铆接	鉚接	riveting
冒口	冒口(鑄造)	riser, gate riser
眉毛板	窗楣	eye brow, brow
门槽	門槽	dock gate channel
门库	閘門室	gate chamber
孟克力矩	孟克力矩	Munk moment
迷宫式汽封	迷宮迫緊,曲徑填封	labyrinth packing
免除处所	免[除]丈[量]空間	excluded spaces
免除证书	豁免證書	Exemption Certificate
面板	面板	face plate
面板切斜连接	短角鐵連接	clip connection
面层	表塗層	topcoat

祖国大陆名	台湾地区名	英文名
面积比	面積比	area ratio
面罩	面罩	helmet, mask, helmet shield
灭火器	滅火器	fire extinguishers, extinguisher
明轮	明輪	paddle wheel, feathering paddle wheel
模锻	模鍛	die forging
模块	模組,模件	module
模样	模型	pattern
膜式水冷壁	薄膜壁	membrane wall
摩擦焊	摩擦焊	friction welding
摩擦式离合器	摩阻離合器	friction clutch
摩擦阻力	摩擦阻力	frictional resistance
摩擦阻尼	摩擦阻尼	frictional damping
摩振腐蚀	磨[耗腐]蝕	fretting corrosion
磨合	適配運轉	running-in
末端卸扣	端接環	end shackle
末叶片	最後葉片	final blade
莫氏信号灯	莫氏信號燈	Morse signal light
母材	母材	base metal, parent metal
母线	母線	generating line
母子式渔船		mother-ship with fishing dory
木船	木船	wooden vessel, wooden ship
木工间	木工庫	carpenter store

N

祖国大陆名	台湾地区名	英文名
耐波性	耐海性,耐波性能	seakeeping qualities, seakeeping perform-ance
耐波性衡准	耐海性準則	criteria of seakeeping qualities
耐波性试验	耐海性試驗	seakeeping test
耐波性试验水池	耐海性試驗水槽	seakeeping tank
耐潮绝缘材料	抗濕絕緣材料	moisture resistant insulating material
耐火电缆	抗燃電纜	fire resisting cables
耐火度	耐火度,耐火性	refractoriness
耐火救生艇	防火救生艇	fire-resistant lifeboat, fire protected life-boat
耐火黏土	耐火泥,火泥,燒磨土	fireclay, chamot, chamotte
耐火砖	耐火磚,火磚,燒磨土磚	firebrick, chamotte brick

祖国大陆名	台湾地区名	英文名
耐久[性]试验	持久試驗	endurance test
耐磨性	耐磨耗性	wear resistance
耐压壳体	壓力殼(潛艇)	pressure hull
耐压壳体进水		flooding of pressure hull
挠性转子	柔性轉子	flexible rotor
内侧轴系	内軸	inner shafting, inner shaft
内底板	内底板	inner bottom plating
内底边板	二重底緣板	margin plate
内底横骨	副肋骨	reversed frame
内底纵骨	内底縱材	inner bottom longitudinal
内河船	内河船,内陸水域船	inland vessel
内倾	船舷内傾	tumble home
内燃机动力装置	内燃機動力設備	internal combustion engine power plant
内舾装		accommodation outfitting
内效率	内效率	internal efficiency
内旋	内[向旋]轉(螺槳)	inward turning, inboard turning
内应力	内應力	internal stress
泥泵	挖泥泵	dredge pump
泥驳	開底泥駁[船]	hopper barge
泥舱容积	泥艙容量	hopper capacity
泥斗	斗	bucket
泥箱	泥箱	mud box
逆火	逆火	back fire, flashback
逆螺线试验	逆蝸旋試驗	reverse spiral test
黏度	黏度	viscosity
黏性阻力	黏性阻力	viscous resistance
黏压阻力	黏性壓差阻力	viscous pressure resistance
凝聚	凝聚	coagulation, cohesion
凝汽器	冷凝器,凝結器	condenser
[凝汽器]冷却面积	冷卻面	condenser cooling surface, cooling surface
凝汽式汽轮机	凝汽渦輪機	condensing turbine
凝水泵	凝水泵	condensate pump
扭矩	扭矩,轉矩	torque
扭曲	扭曲	twisting
扭转振动	扭轉振動	torsional vibration
农用船		agricultural vessel

O

祖国大陆名	台湾地区名	英 文 名
耦合	偶合	coupling
耦合运动	偶合運動	coupling motion

P

祖国大陆名	台湾地区名	英 文 名
耙吸挖泥船		trailing suction hopper dredger
排泥管接岸装置	排泥管接岸裝置	shore connecting plant
排泥设备	排泥設施	soil discharging facility
排气槽	通氣孔,通氣口	air vent
排气道	排氣導管	exhaust duct
排气风机	排氣機,抽風機	exhaust fan
排气行程	排氣衝程	exhaust stroke
排气装置	排氣裝置	exhaust unit
排汽	排汽	exhaust steam
排汽系统	排汽系統	exhaust steam system, exhaust system
排水量	排水量	displacement
排水型船	排水型船	displacement ship
排污	沖放,放水	blow down
排污阀	底部吹泄閥,沖放閥	blowdown valve
排油监控系统	泄油監控系統	oil discharge monitoring system, oil discharge monitoring and control system
盘线装置	捲線機	line winder
旁桁材	側縱梁	side girder
旁内龙骨	側內龍骨	side keelson
旁通阀	旁通閥	by-pass valve
旁通调节	旁通調節	by-pass control, by-pass governing
抛光	擦光	buffing
抛落式救生艇		free fall lifeboat
抛锚试验	錨試驗	anchoring trial, anchor test
泡沫灭火器	泡沫滅火器	foam fire extinguisher, foamite extinguisher
泡沫灭火系统	泡沫滅火系統	foam fire extinguishing system

祖 国 大 陆 名	台 湾 地 区 名	英 文 名
炮架	炮架	carriage, gun rest
炮身窗护板	炮擋,炮圍	gun port shield, gun shield
配电屏	饋電盤	feeder panel
配气定时	閥動定時	valve timing
配汽机构	配汽機構	steam distribution device
配重	配重	mass-balance weight, counter weight
喷淋设备	噴水系統	sprinkler system
喷气推进	噴射推進	jet propulsion
喷气推进船	噴氣推進艇	air jet propelled boat
喷气推进器	噴射推進器	jet propulsion unit, jet propeller
喷枪	噴槍	spray gun
喷洒装置	噴灑裝置,撒佈器	sprinkler
喷砂	噴砂[除銹]	sand blasting, sand blast
喷射泵	噴射泵	jet dredge pump, jet pump
喷射泵挖泥船	噴射式挖泥船	jet ejector dredger
喷射处理	噴砂處理	blasting
喷射过渡	噴灑狀傳遞	spray transfer
喷水推进	噴水推進	water jet propulsion
喷水推进船	噴水推進船,噴射推進船	water jet [propelled] boat, hydrojet propelled ship, jet propulsion ship
喷水推进器	噴水推進器	waterjet propulsor
喷水推进燃气轮机	噴水推進燃氣渦輪機	water-jet propulsion gas turbine
喷涂	噴塗	spraying, spray painting
喷丸	噴[射鋼]珠除銹,珠[粒噴]擊	shot blasting
喷丸机	噴珠除銹機	shot blaster chamber, shotblaster
喷油压力	噴射壓力	injection pressure
喷嘴	噴嘴,噴口	nozzle
喷嘴室	噴嘴箱	nozzle box
碰垫	碰墊	fender
砰击	波擊	slamming
砰击载荷	波擊負荷	slamming load
膨胀行程	膨脹衝程	expansion stroke
膨胀[压]比	膨脹比	expansion [pressure] ratio
坯料	坯料	billet
疲劳	疲勞	fatigue
片式离合器	盤形離合器	disc clutch
片体	[雙體船的]半[船]體	demihull

祖 国 大 陆 名	台 湾 地 区 名	英 文 名
片状空化	片狀空化	sheet cavitation
偏心度	偏心度,偏心距	eccentricity
漂浮式下水	漂浮下水	floating launching
漂浮烟雾信号	浮[式發]煙信號	buoyant smoke signal
漂心	浮面中心	center of floatation
漂心纵向坐标	縱向浮[力中]心	longitudinal center of floatation
撇缆绳	撇纜繩,引纜繩	heaving line, hauling line
品质因数	品質因數	quality factor
平板舵	平板舵,單板舵	single plate rudder
平板龙骨	平板龍骨	plate keel
平舱	整平(散裝貨)	trimming
平焊	平焊,平焊接	flat position welding, flat welding, flush weld
平衡舵	平衡舵	balanced rudder
平衡活塞	均壓活塞	dummy piston
平衡通风	平衡通風	balanced draft
平甲板船	平甲板船	flush deck ship, flush deck vessel
平均吃水	平均吃水	mean draft
平均海平面	平均水位	mean sea level
平均修复时间	平均修復時間	mean repair time
平均叶宽	平均葉寬	mean blade width
平均叶宽比	平均葉寬比	mean blade width ratio
平均有效压力	制動平均有效壓力	brake mean effective pressure
平面舱壁	平面艙壁	plane bulkhead
平面分段		flat section
平面运动机构试验	平面運動機構試驗	planar motion mechanism test, PMM test
平台罗经	平台羅經	stabilized gyrocompass
平行中体	平行舯體	parallel middle body
平旋推进器	擺線推進器,垂直翼螺槳	cycloidal propeller, Voith-Schneider propeller
屏蔽层	簾幕	screen
坡口角度	開槽角	groove angle
坡口面	開槽面	groove face
坡口面角度	斜角	bevel angle
破冰船	破冰船	icebreaker
破波阻力	碎波阻力	wave breaking resistance
破舱水线	浸水吃水線	flood waterline, flooded waterline
破舱稳性	破損穩度	damaged stability

祖国大陆名	台湾地区名	英 文 名
破坏性试验	破壞性試驗	destructive test, destruction test
普遍腐蚀	全面腐蝕	general corrosion

Q

祖国大陆名	台湾地区名	英 文 名
脐带	[潛水具]供應連系管，臍索	umbilical
旗箱	旗箱	flag chest, flag locker
鳍	鳍板	fin
起动[过程]电流	起動電流	starting current
起动[过程]转矩	起動轉矩，起動扭矩	starting torque
起动空气分配器	起動空氣分配器	starting air distributor
起动空气系统	起動空氣系統	starting air system
起动器	起動器	starter
起动试验	起動試驗	starting test
起动压力	起動壓力	starting pressure
起动装置	起動設施	starting device
起货机平台	絞機台	winch platform
起货绞车	吊貨絞機	cargo winch
起货设备	貨物升降機	cargo lift equipment, cargo lift
起货索具	吊桿索具	derrick ringging
起居舱室	起居艙[設施]，住艙	accommodation, living quarter
起居甲板	住艙甲板，起居艙甲板	accommodation deck
起落绞车	起吊絞機	hoisting winch
起锚船	拋錨艇	anchor boat
起锚机	錨機	windlass
起锚绞盘	起錨絞盤	anchor capstan
起皮	剝離	peeling
起艇绞车	小艇絞車	boat winch
起网机	起網機	net winch, net hauler
起网机组	起網系統	net hauling system
起重船	起重[工作]船	floating crane, crane ship, derrick barge
起重桅	吊桿桅	derrick mast
起重柱	吊桿柱	derrick post
起皱	皺紋	wrinkles
起座压力	泄氣壓力（安全閥）	popping pressure
气道	導氣管	air duct

祖国大陆名	台湾地区名	英 文 名
气笛	汽笛,氣笛,號笛	air horn, siren, syren
气垫	氣墊	cushion, air cushion
气垫船	氣墊船	air cushion vessel, ACV, air cushion vehicle
气缸	氣缸	cylinder
气缸盖	缸頭,缸蓋	cylinder head, cylinder cover
气缸容积	氣缸容積	working medium volume, cylinder volume
气缸套	缸襯[套],缸套	cylinder liner, cylinder jacket
气缸体	缸體	cylinder block
气缸直径	缸内徑	cylinder bore
气割	焰割	gas cutting
气焊	氣焊	gas welding
气孔	氣孔	blowhole
气力提升泵	吸泥泵	air lift mud pump
气门升程	閥升程	valve lift
气体保护[电弧]焊	氣體遮護金屬電[弧]焊[接]	gas shielded arc welding, gas metal arc welding, GMAW
气体浓度测量仪	氣體濃度測量儀	gas concentration measurement instrument
气胀救生筏	充氣式救生筏	inflatable liferaft
弃链器	釋纜器	cable releaser
汽封	汽封	steam seal gland, steam seal
汽封系统	渦輪汽封系統	turbine steam seal system
汽缸	汽缸	cylinder
汽耗量	蒸汽消耗量	steam consumption
汽耗率	汽耗率	steam rate, specific steam consumption
汽化油灶	汽化油竈	vaporizing oil range
汽轮鼓风机	渦輪鼓風機	turbo-blower
汽轮机	蒸汽渦輪機	steam turbine
汽轮机船	汽船	steam turbine ship
[汽轮机]主轴	主軸	main shaft
汽水共腾	汽水共騰	priming
汽水阻力	壓力降	pressure drop
汽油机	汽油機,汽油引擎	gasoline engine, petrol engine
千斤索	俯仰頂索(吊桿),跨索	topping lift, span rope
千斤索具	跨索具	span ringging
千斤座	[桅頂]俯仰滑車座	topping bracket
钎焊	硬焊	brazing
牵索	牽索,控索	guy

祖 国 大 陆 名	台 湾 地 区 名	英 文 名
牵引链条	制動鏈(下水)	drag chain
前端壁	前端壁	front bulkhead
前体	艏部	fore body
前桅	前桅	fore mast
潜水服	潛水服,潛水衣	diving suit
潜水工作船,潜水支援船	潛水工作船,潛水支援船	diving support vessel, diving boat
潜水器	潛水器,潛水載具	submersible, submersible vehicle
潜水器母船	潛水器母船	mother ship of submersible
潜[水]艇	潛[水]艇,潛艦	submarine
潜水支援船(＝潜水工作船)		
潜水钟	潛水鍾	diving bell
潜水装具	潛水器具	diving apparatus
潜艇救生船	潛艇救難艦	submarine rescue ship, submarine rescue vessel
潜艇鱼雷发射装置	魚雷發射管	torpedo launcher for submarines, torpedo tube
潜望镜	潛望鏡	periscope
浅水效应	淺水效應	shallow water effect
强横梁	強力梁,大梁	web beam
强浸蚀	浸洗	pickling
强肋骨	大肋骨	web frame
强力甲板	強度甲板	strength deck
强胸横梁	抗拍梁	panting beam
强胸结构	抗拍結構	panting arrangement
强制循环锅炉	強制循環鍋爐	forced circulation boiler, assisted circulation boiler
敲渣锤	鏨鍾	chipping hammer
桥规	橋[形]規	bridge gauge
桥楼	橋樓,駕駛台	bridge
翘度	翹度,[螺葉]後傾	set-back, wash-back
切割顺序	切割順序	cutting sequence
切割线	切割線	line of cut, cutting line
切削速度	切削速度	cutting speed
切斜端	切角端	snip end
氢脆	氢脆化	hydrogen embrittlement
氢氧焊	氢氧焊接	oxy-hydrogen welding

祖 国 大 陆 名	台 湾 地 区 名	英 文 名
氢致破裂	氢脆裂	hydrogen induced cracking
轻舱壁	隔間艙壁,屏隔艙壁	partition bulkhead, screen bulkhead
轻柴油输送泵	輕柴油輸送泵	light diesel oil transfer pump
轻合金船	輕合金船	light alloy ship
轻型吊杆	輕型吊桿	light derrick boom
轻型护卫舰	輕型巡防艦	corvette
轻型肋板	減輕孔肋板	lightened floor
倾斜试验	傾側試驗	inclining test, inclining experiment
清铲	鑿平	chipping
清根	背縫鑿淨,背縫鑿平	back chipping
清洁压载舱操作手册	清潔壓艙水操作手冊	clean ballast tank operation manual
清洁压载水	清潔壓艙水	clean ballast
清洗试验	清洗試驗	washing test
清洗装置	清洗裝置	washing equipment
球鼻艏	球[形]艏	bulbous bow
球形接头	球接頭	ball joint
球形艉	球形艉	bulbous stern
区配电板	分段配電板,區配電板	section board
曲柄半径	曲柄半徑	crank radius
曲面分段		curved section
曲轴	曲[柄]軸	crankshaft
曲轴箱	曲柄軸箱	crankcase
曲轴箱防爆门	曲柄軸箱防爆門	crankcase explosion proof door, crank- case explosion relief door
曲轴箱扫气	曲柄軸箱掃氣	crankcase scavenging
驱气	清除[有害]氣體	gas-freeing
驱气系统	貨油艙清除油氣裝置	gas freeing, cargo oil tank gas-freeing in- stallation
驱逐舰	驅逐艦	destroyer
去极化	去極化作用	depolarization
去流段	艉出水段	run
去流段长	艉出水段長	length of run
去应力退火	應力消除熱處理	stress relieving, stress relief heat treatment
全程序调速器	可變調速器	variable speed governor
全船报警装置		general alarm
全垫升气垫船	氣墊船	hovercraft
全封闭救生艇	全圍蔽救生艇	totally enclosed lifeboat
全负荷	全負荷	full load

祖国大陆名	台湾地区名	英　文　名
全回转推进器	全向螺槳	all-direction propeller
全潜船	全潛船	underwater ship
全球定位系统	全球[衛星]定位系統	global positioning system, GPS
全球海上遇险和安全系统	全球海上遇險及安全系統	global maritime distress and/or safety system, GMDSS
全燃联合动力装置	組合燃氣渦輪機	combined gas turbine and/or gas turbine power plant, COGAG, COGOG
裙板	裙板	skirt plate
群速度	群速	group velocity

R

祖国大陆名	台湾地区名	英　文　名
燃点	燃點	fire point, inflammable point
燃料系数	耗油係數	fuel coefficient
燃料消耗试验	燃料消耗試驗	fuel consumption test
燃料消耗量	燃料消耗量	fuel consumption
燃料油输送泵	燃油輸送泵	fuel oil transfer pump
燃煤锅炉	燃煤鍋爐	coal firing boiler, coal burning boiler
燃气发生器	氣體發生器	gas generator
燃气轮机	燃氣渦輪機	gas turbine
燃气轮机船	燃氣渦輪機船	gas turbine ship
燃气轮机动力装置		gas turbine power plant
[燃烧]过量空气系数	過量空氣因數	excess air ratio, excess air factor
燃烧率	燃燒率	combustion rate
燃烧器	燃燒器	burner, combustor
燃烧器喷口	燃燒器嘴	burner port, burnertip tip
燃烧室	燃燒室	combustion chamber, fire combustion chamber
燃烧室外壳	燃燒室外殼	combustor outer casing
燃烧效率	燃燒效率	combustion efficiency
燃油舱	燃油艙	fuel oil tank
燃油沉淀柜	燃油沉澱櫃	fuel oil settling tank
燃油废气组合式锅炉	廢氣燃油複合式鍋爐	composite oil-exhaust gas fired boiler, composite oil-fired/exhaust gas boiler
燃油柜	燃油櫃	fuel oil tank
燃油滤器	燃油過濾器	fuel oil filter
燃油日用柜	燃油日用櫃	fuel oil daily tank

祖 国 大 陆 名	台 湾 地 区 名	英 文 名
燃油系统	燃油系统	fuel oil system
燃油消耗量	燃油消耗量	fuel consumption
[燃]油[消]耗率	單位耗油量	specific fuel consumption
燃油泄放系统	燃油泄放系統	fuel oil drain system
扰动力	激振力	exciting force
绕射力	繞射[波]力	diffraction force
热成形	加熱成型	hot forming
热冲击	熱衝擊	thermal shock
热处理	熱處理	heat treatment
热脆	熱脆性	hot shortness
热镀锌	鍍鋅	galvanizing, galvanization
热负荷	熱負載	thermal load
热耗率	耗熱率	heat rate
热剂焊	發熱焊接,鋁熱焊接	thermit welding
热交换器	熱交換器	heat exchanger
热节	高熱點	hot spot
热井	熱井,串列柜	cascade tank, hot well
热裂	熱裂,高溫龜裂	hot tearing, hot cracking
热面点火	表面點火	surface ignition
热疲劳	熱疲勞	thermal fatigue
热平衡	熱平衡	heat balance
热气机动力装置	史特靈引擎動力設備	stirling engine power plant
热水锅炉	熱水鍋爐	hot water boiler
热损失	熱損失	heat loss
热态起动	熱起動	hot starting
热效率	熱效率	thermal efficiency
热应力	熱應力	thermal stresses
热影响区	熱影響區[域](焊接)	heat-affected zone
热轧氧化皮	軋鋼鱗片	mill scale
热装	收縮配合,紅套	shrinkage fitting, shrinkage fit
热阻	熱阻	thermal resistance
人工控制	人工控制,手動控制	manual control
人工越控装置	人工越控裝置	manual override system
人工智能	人工智能	artificial intelligence
人孔	人孔	man hole
人孔盖	人孔蓋	manhole cover
人力应急操舵试验	人力緊急操舵試驗	manual emergency steering test
人力应急起锚试验	人力緊急起錨試驗	manual emergency anchoring test

祖国大陆名	台湾地区名	英 文 名
人字式坞门	人字式塢門	mitre caisson, two-gate caisson
人字桅	人字桅,雙腳桅	bipod mast
韧性	韌性,韌度	toughness
日用淡水泵	日用淡水泵	daily service fresh water pump
日用燃油泵	日用燃油泵	daily service fuel oil pump
容积曲线	容積曲線	capacity curve
溶剂	溶劑	solvent
溶蚀	沖蝕,潰蝕	erosion
熔池	熔[化]池(焊接)	molten pool
熔滴	熔滴(電焊)	globule
熔断器	熔斷器	fuse
熔敷金属	堆積金屬,溶著金屬 (焊接)	deposited metal, deposition metal
熔敷顺序	組合順序(焊接)	build up sequence
熔敷速度	溶著速度	deposition rate
熔敷效率	澱積效率,堆積效率,溶 著效率(焊接)	deposition efficiency
熔焊	熔接	fusion welding
熔核	焊塊	nugget
熔深	熔解深度	penetration, depth of fusion
柔韧性	柔性,柔度	flexibility
柔性制造单元		flexible manufacturing cells, FMC
柔性制造系统	彈性製造系統	flexible manufacturing system, FMS
蠕变	潛變(材料)	creep
乳化	乳化	emulsification
乳化燃料	乳化燃料	emulsified fuel
入级	入級,船級	classification
入级检验	船級檢驗	classification survey
入级证书	船級證書	classification certificate
入射角	入射角,攻角	angle of incidence, incident angle
软电缆	柔性電纜,柔性纜	flexible cable
软钎焊	軟焊	soldering
软钎料	軟焊料,焊錫	solder
软梯	軟梯,繩梯	rope ladder

S

祖 国 大 陆 名	台 湾 地 区 名	英 文 名
洒水短管	灑水短管	spraying tube
塞焊	塞孔焊接	plug welding
赛艇	競賽艇	racing boat, runner boat
三角眼板	三角眼板	triangular plate
三脚桅	三腳桅	tripod mast
三联泵	三缸泵	triplex pump
三体艇	三[胴]體船	trimaran
散货船	散装貨船	bulk carrier, bulk cargo ship
散射锚泊系统	分散錨泊	spread anchoring system
散射系泊定位钻井船	分散系泊定位鑽探船	spread moored drilling ship
散装货物	散裝貨	bulk cargo
扫舱泵	殘油泵,收艙泵	stripping pump
扫舱总管	收艙總管	stripping main line
扫雷舰艇	掃雷艇	minesweeper
扫雷具	掃雷具	sweep
扫描	掃描	sweep
扫气箱	驅氣接收器	scavenging air receiver
扫气效率	驅氣效率	scavenging efficiency
砂箱墩	砂箱墩,砂墩	block with sand box, sand block
舢板	舢板	sampan
闪点	閃點,引火點	flash point
闪光	閃光	flashing, flash
闪光对焊	閃電對頭焊接	flash butt welding
商船	商船	merchant ship, commercial ship
上层建筑	上層建築,船樓建築	superstructure
上层建筑甲板	船樓甲板	superstructure deck
上层建筑整体吊装		lifting and mounting complete superstruc-ture
上舵承	舵承上部,上舵承	rudder carrier, upper bearer
上浮	上浮	surfacing
上甲板	上甲板	upper deck
上排污阀	液面吹泄閥	surface blowdown valve, surface blow off valve

祖国大陆名	台湾地区名	英 文 名
上升管	升導管	riser
上桅	上桅	topgallant mast
上止点	上死點	top dead center
梢厚	葉尖厚	blade tip thickness
梢隙	葉尖間隙	blade tip clearance
设计吃水	設計吃水	designed draft
设计航速	設計船速	design speed
设计基准	設計基準(準則)	design datum, design norm
设计排水量	設計排水量	designed displacement
设计水线	設計水線	designed waterline
射程	射程	range
射频干扰	無線電頻率干擾	radio interference, radio frequency inter-ference
射水抽水器	射水抽水器	water ejector
X 射线	X 射線	X-radiation, X-ray
射线透照检查	放射線檢查	radiographic inspection
伸缩接管	伸縮接合	expansion joint
伸缩接头	伸縮接頭	expansion joint
伸张面积	伸展面積(螺槳)	expanded area
深舱	深艙	deep tank
深舱肋骨	深肋骨	deep tank frame, deep frame
深潜救生艇	深潛救難艇	deep submersible rescue vehicle, deep submergence rescue vehicle, DSRV
深潜器	深潛器	deep diving submersible, bathyscaphe
深潜系统	深海潛水系統	deep diving system
深熔焊	深透焊接	deep penetration welding
深水炸弹	深水炸彈	depth charge
甚长波通信		very long wave communication
渗碳	滲碳法	carburizing, carburization
渗锌	滲鋅	sheradizing
升船机	升船機	shiplift, ship elevator
升降式舱口盖	升降式艙口蓋	lift hatchcover
升力阻尼	升力阻尼	lift effect damping
生产测试船		production test ship
生产储油船		production storage tanker
生存能力	存活性	survivability
生活污水	穢水,[生活]污水	sewage
生活污水储存柜	生活污水儲存櫃	sewage holding tank

祖 国 大 陆 名	台 湾 地 区 名	英 文 名
生活污水处理装置	穢水處理裝置	sewage treatment plant
生铁	生鐵	pig iron
声波	聲波,音波	acoustic wave
声力电话机	聲力電話	sound powered telephone
声呐导流罩涂料	聲納罩塗料	sonar dome paint, sonar dome coating
声呐电缆绞车	聲納電纜絞車	sonar cable winch
声频,音频	聲頻	audio frequency, acoustic frequency
声响信号	音響信號,聽覺信號	audible signal
声压	聲壓	sound pressure
绳扣	吊索,吊鏈	sling
剩磁	剩磁	residual magnetism
剩余阻力	剩餘阻力	residual resistance, residuary resistance
失控灯	[船]操縱失靈號燈	not under command light
失效率	失效率,故障率	failure rate
失效模式	故障模式	failure mode
失效模式、效应及后果 分析	故障模式及影響分析	failure mode effect and criticality analysis, failure mode and effect analysis, FMEA
施救浮索	營救浮環	buoyant rescue quoit
湿舱	濕艙	wet chamber
湿面积	浸水面	wetted surface
湿喷砂除锈	濕噴砂除銹	wet sand blasting
湿碰湿	濕式積層	wet on wet
湿式潜水器	濕式潛水器	wet submersible
湿态强度	濕態強度	wet strength
十字带缆桩	十字形系樁	cross bitt
十字接头	十字接頭	cross shaped joint, cross joint
时间继电器	延時繼電器	time-delay relay
时间间隔	時間延遲,時滯	time lag
识别板	銘牌	name plate
实尺放样	實尺放樣	full scale lofting
实肋板	實體肋板	solid floor
实效伴流	有效跡流	effective wake
实效重量	虚質量	virtual mass
实心轴	實心軸	solid shaft
实芯绝缘电缆	實芯絕緣電纜	solid dielectric cables
史密斯效应	史密斯效應	Smith effect
使用寿命	有效壽命	useful life
示功阀	指示閥	indicator cock, indicator valve

祖国大陆名	台湾地区名	英 文 名
示功图	示功圖	indicator diagram
试航速度	試航船速	trial speed
试件,试样	試件,試樣	specimen, test specimen
试样(=试件)		
适伴流螺旋桨	適跡[流]螺槳	wake-adapted propeller
适航性	適航性	seaworthiness
适应控制	適應控制	adaptive control
收缩	收縮	shrinkage
收缩余量	收縮裕度	shrinkage allowance
手持火焰信号(=手持烟火信号)		
手持烟火信号,手持火焰信号	手持火焰信號	hand fire signal, hand flare
手动泵	手搖泵	hand pump
手动火灾警报器,手动火灾报警装置	手動火災警報系統	manual fire alarm sounder, manual fire alarm system
手动火灾报警装置(=手动火灾警报器)		
手工除锈	手工敲劗除銹	handtool cleaning, hand cleaning
手工焊	人工焊接	manual welding
手弧焊	金屬被覆電弧焊	shielded metal arc welding, SMAW
手提式灭火器	輕便滅火器	portable fire extinguisher
艏部结构	艏結構	stem structure, bow construction
艏沉	艏俯	dipping
艏吃水	艏吃水	fore draft, draught forward, forward draft, forward draught
艏垂线	艏垂標	forward perpendicular, fore perpendicular
艏舵	艏舵	bow rudder
艏尖舱	艏尖艙,前尖艙	fore peak, forepeak tank
艏尖舱泵	艏尖艙手動泵	forepeak pump, head pump
艏楼	艏樓	forecastle
艏落	艏落	dropping
艏锚	艏錨,大錨,主錨	bow anchor
艏锚绞车		bow position winch
艏门		bow door
艏翘	艏部上翹變形	cocking up of forebody
艏倾	艏俯	trim by bow
艏斜浪	艏側浪	bow sea

祖 国 大 陆 名	台 湾 地 区 名	英 文 名
艏摇	[艏艉]平擺	yawing
艏支架	艏支架,艏托台(下水)	fore poppet
艏踵	艏材跟部	forefoot
艏柱	艏柱,艏材	stem
受热面	受熱面	heating surface
枢心	轉向軸心	pivoting point
疏水	排泄	drain
疏水泵	污水泵	drainage pump, drain pump
疏水阀	排泄閥	drain valve
疏水管	排泄管	drain pipe
疏水箱	排泄櫃	drain tank
输油平台	輸油平台	transfer platform
竖桁	豎桁	vertical girder
数控平面运动拖车	數控平面運動拖車	computer controlled planar motion carriage, computerized planar motion carriage
数学放样	數學放樣	mathematical lofting
刷涂	毛刷塗裝	brush painting
衰减	衰減	attenuation
双板舱壁	雙層艙壁	double plate bulkhead, double bulkhead
双层底	[二]重底	double bottom
双层底舱	[二]重底艙	double bottom tank
双杆吊货装置	雙吊桿作業系統	union purchase system
双环盲板法兰	眼鏡型盲凸緣	spectacle blind flange, spectacle blank flange
双截门	雙截門	double-leaf door
双联泵	雙缸泵	duplex pump
双流汽轮机	雙流渦輪機	double-flow steam turbine, divided flow turbine, double flow turbine
双路供电	雙路供電,雙重供電	two circuit feeding, duplicate supply
双面焊	雙面焊接接合	welding by both side
双面坡口	雙面槽	double groove
双壳船	雙殼船	double-skin ship, double hulled ship
双燃料发动机	雙燃料引擎	dual-fuel engine
双体船	雙[胴]體船	catamaran, twin-hull ship
双拖渔船	雙拖[網]漁船	bull trawler, two boat trawler
双艉鳍	雙艉鳍	twin-skeg
双向桁架	雙向桁架	two-direction truss

祖国大陆名	台湾地区名	英 文 名
双支承舵	雙支承舵	double bearing rudder
双作用泵	雙動泵	double acting pump
水封式漏水口	水封式排水口	water seal scupper
水管锅炉	水管鍋爐	water tube boiler
水雷	水雷	mine
水雷式代换水舱	水雷補重櫃（潛艇）	mine compensating tank
水冷	水冷卻	water cooling
水密舱壁	水密艙壁	watertight bulkhead
水密分舱	水密艙區劃分	watertight subdivision
水密肋板	水密肋板	watertight floor
水密门	水密門	watertight door
水面高程	水面高程	water level elevation
水面舰船	水面船	surface ship
水平弯曲振动	水平彎曲振動	horizontal flexural vibration
水平指示灯	海平面指示燈	horizon indicating lamp
水平肘板	水平肘板	horizontal bracket
水上航速	水面速［率］	surface speed
水上合拢		afloat joining ship sections
水声调查船	水聲調查船	underwater acoustic research vessel, underwater acoustic research ship
水声反声材料	水下反音材	underwater acoustic reflection material
水声透声材料	水下透音材	underwater acoustic transmission material
水声吸声材料	水下吸音材	underwater acoustic absorption material
水听器,接收换能器		receiving transducer
水位	水位	water level
水文调查船		hydrological survey vessel, hydrological survey ship
水下爆炸试验水池	水下爆炸試驗水槽	underwater explosion tank
水下侧面积	水下側［向］面積	lateral underwater area
水下逗留时间	水下續航力	submerged endurance
水下焊接	水中焊接	underwater welding
水下监听站	水下監測站	underwater monitoring station
水下胶黏剂	水下膠黏劑	underwater adhesive
水下居住舱	水下居住艙	underwater habitat
水下黏合	水下黏合	underwater adhesion
水下切割	水中切割	underwater cutting
水下生产系统	海底石油生產系統	subsea production system
水下游览船	水下遊覽船	tourist submersible, underwater sightsee-

祖国大陆名	台湾地区名	英 文 名
		ing-boat
水下作业灯	水下作業燈	submerged lamp
水下作业机械	水下作業機械	underwater operating machine
水下作业站	水下工作站	underwater working station
水线	水線	waterline
水线长	水線長度	waterline length
水线面	水線面	water plane
水线区腐蚀	水線區腐蝕	waterline zone corrosion, boottopping corrosion
水线涂料	水線漆	boottopping paint
水压试验	[静]水壓[力]試驗	hydrostatic test
水压载	水壓載	water ballast
水翼艇	水翼船	hydrofoil craft, hydrofoil boat
水阻	水阻力	water resistance
瞬态	暫態,過度狀態,瞬時狀態	transient state, transient mode
瞬态波试验	暫態波試驗	transient wave test
瞬态电流	暫態電流	transient current
死区	死空間	dead metal region, dead space
四冲程内燃机	四衝程引擎	four-stroke engine, four cycle engine
四冲程循环	四衝程循環	four-stroke cycle, four cycle
送丝机构	供線機	wire feeder
送丝装置	供[焊]線裝置	wire drive feed unit
送网管	送網管	net carrying pipe
苏伊士运河吨位	蘇彝士運河噸位	Suez canal tonnage
苏伊士运河探照灯	蘇彝士運河探照燈	Suez canal search light
苏伊士运河专用吨位证书	蘇彝士運河噸位證書	Suez Canal Special Tonnage Certificate
速闭舱盖	快速開閉艙蓋	quick-closing hatchcover, quick acting hatchcover
速度级	速度級	velocity stage
速度系数	速度係數	velocity coefficient
速潜	緊急下潛	quick diving
塑性	塑性	plasticity
酸度	酸度	acidity
酸洗	酸洗	pickling
随边	殿緣	trailing edge
随浪	順浪,從浪	following sea

祖 国 大 陆 名	台 湾 地 区 名	英 文 名
隧道顶线	隧［道］頂線	tunnel top line
隧道艉	隧［道］艉	tunnel stern
损管设备室		damage control equipment room
损伤	損傷	damage
损失水线面面积	損失水線面面積	lost waterplane area
缩尺	收縮尺	shrinkage rule
缩尺比	尺度比	scale ratio
索具	索具	rigging

T

祖 国 大 陆 名	台 湾 地 区 名	英 文 名
踏步	踏板	foot steps
胎架	組合模架	assembly jig
台架试验	試台試驗	bench test
钛钙型焊条	鹼基氧化鈦系焊條	lime titania type electrode
弹性联轴器	彈性聯軸器	resilient shaft coupling, elastic coupling
碳弧焊	碳［極電］弧熔焊	carbon arc welding
碳纤维增强塑料	碳纖維強化塑膠	carbon fiber reinforced plastics
镗孔	搪孔	boring
淘空	海泥流失（鑽油井）	scouring
套管连接	套筒接合	sleeve joint
套环	牛眼圈,纜索嵌環	thimble
套筒联轴器	套筒聯結器	sleeve coupling
特殊区域	特別海域	special area
特性	特性	characteristic
特性阻抗	特性阻抗	characteristic impedance
梯道布置	梯道與信道佈置	stairway and passage way arrangement
天窗	天窗	skylight
天幕	天遮	awning
天幕帘	天遮簾［幕］	awning curtains
天幕索	天遮索	awning rope
天气船	氣象觀測船	weather ship, weather vessel
天桥	天橋,連橋	connecting bridge, cat walk
天文导航	天文導航	celestial navigation
天线	天線	antenna
填充物	填料,填充劑	filler
调风器	調氣器	register

祖 国 大 陆 名	台 湾 地 区 名	英 文 名
调伸长度	懸伸	overhang
调速电动机	調速馬達	adjustable-speed motor
调速器	調速器	governor
调速装置	速度控制器	speed control device, speed controller
调谐因子	調諧因子	tuning factor
调质	淬火及回火	quenching and tempering
跳板	跳板,著陸板	ramp
跳板门	登陸舌門,登陸跳板	ramp door
贴合面	接合面	faying surface
铁粉焊条	鐵粉系焊條	iron powder type electrode
铁梨木	鐵梨木	lignumvitae
停船冲程	正[慣性]滑行距離	head reach
停船试验	停船性能試驗	stopping test
艇	小艇,船艇	boat, craft
艇吊钩	吊鈎	lifting hook
艇甲板	小艇甲板	boat deck
艇座	小艇座	boat chock
通道	通道	passage, alleyway
通风机室	風扇室	fan room
通风阻力	通風損失	draft loss
通海阀	通海閥,海水吸入閥	sea suction valve
通海旋塞	海底旋塞	seacock
同步电动机	同步馬達	synchronous motor
同步电机	同步機	synchronous machine
同步发电机	同步發電機	synchronous generator
同步转速	同步轉速	synchronous speed
同轴电缆	同軸電纜	coaxial cable
投影面积	投影面積	projected area
投影面积比	投影面積比	projected area ratio
透气孔	氣孔	air hole
凸焊	凸出焊接	projection welding
凸轮间隙	凸輪間隙	cam clearance
凸轮轴	凸輪軸	cam shaft
凸模	沖頭	punch
凸缘联轴器	凸緣連接器	flanged coupling
涂底漆	刷底漆	priming
涂料施工(=涂装)		
涂膜	塗膜	film

祖 国 大 陆 名	台 湾 地 区 名	英 文 名
涂装,涂料施工	塗裝,塗漆	painting, coating
推进电机		propulsion electric machine
推进风机	推進風機	propulsive fan
推进机组	推進裝置	propulsion units
推进器	推進器,螺槳	propeller, propulsor
推进器柱	螺槳柱	propeller post
推进系数	推進係數	propulsive coefficient
推进效率	推進效率	propulsive efficiency
推进装置	推進設備	propelling plant
推力减额因数	推[力]減[少]因數	thrust deduction factor
推力块	推力墊,止推墊	thrust pad
推力环	推力[軸]環	thrust collar
推力盘	推力[軸]環	thrust collar
推力系数	推力係數	thrust coefficient
推力仪	推力計	thrust meter
推力轴	推力軸	thrust shaft
推力轴轴承	推力[軸]承,止推[軸]承	thrust bearing, thrust block
推压机构	推壓機構	dipper crowding gear
退磁	消磁,去磁	demagnetization
退火	退火	annealing
托管架	艉托架(布管船)	stinger
托盘	托[貨]板	pallet
拖船	拖船	tug, tugboat, tow boat, towing vessel
拖带灯	拖航燈	towing light
拖钩	拖[纜]鈎	towing hook
拖缆	拖纜,拖索	towing line, towrope
拖缆承架	拖索承梁	towing beam
拖缆绞车	拖纜絞機	towing winch
拖网渔船	拖網漁船	trawler
拖曳弓架	拖纜拱架	towing arch
拖曳航速	拖曳船速	towing speed
拖曳	拖曳	drag
拖曳水池	船模試驗槽	towing tank
拖桩	拖纜樁	towing post
脱气	除氣	deaeration
脱险口	逃生窗口	escape scuttle
脱氧剂	除氧劑,還原劑	deoxidizer

祖 国 大 陆 名	台 湾 地 区 名	英 文 名
陀螺罗经	電羅經	gyrocompass
陀螺罗经室	電羅經室	gyrocompass room
陀螺仪	陀螺儀,回轉儀	gyro, gyroscope
椭圆艉	椭圆艉	elliptical stern

W

祖 国 大 陆 名	台 湾 地 区 名	英 文 名
挖泥船	挖泥船	dredger
外板	船殼板,外板	shell plate
外板展开	外板展開	shell plate development
外侧轴系	外側軸	outer shafting, outer shaft
外加电流阴极保护	外加電流陰極防蝕	impressed current cathodic protection
外廓线	船體側面[圖]	profile
外壳	外殼	enclosure, case
外倾	舷緣外傾	flare
外施电压	外施電壓	applied voltage
外舾装	艙面艤裝	deck outfitting, deck outfit
外斜轴系	[相對於中心面]外斜軸系(多軸系船)	diverging shafting
外形吃水	外型吃水	navigational draft
外旋	外旋	outward turning
弯曲	彎曲	bending
完工检验	完工檢查	final inspection, inspection after construction
完全退火	完全退火	full annealing, dead soft annealing, complete annealing
万向架	萬向吊架,恒平框架	gimbal table, gimbal mounting
万向联轴器	萬向聯結器	universal coupling
网板架	網板架	trawl gallow
危险货物	危險貨物	dangerous cargo, dangerous goods
围蔽处所	圍蔽空間	enclosed spaces
围壁	圍壁	trunk bulkhead
围裙	氣裙,圍裙	skirt
围裙高度	氣裙升高度	skirt depth, skirting rise
围网渔船	圍網漁船,巾著網漁船	seiner
桅	桅	mast
桅灯	桅燈	mast head light, mast light

祖国大陆名	台湾地区名	英 文 名
桅杆涂料	桅桿塗料	mast coating
桅冠	桅頂	truck
桅横杆	帆桁	yard
桅肩	桅上固定具	outrigger
桅设备	桅檣設備	mast and rigging
桅柱	下桅	lower mast
维修	維護,保養	maintenance
维修性	維護性	maintainability
尾滚筒	艉滾筒	stern barrel, stern roll
尾拖渔船	艉拖網漁船	stern trawler
艉部结构	艉段結構	stern structure, stern construction
艉吃水	艉吃水	after draft, after draught
艉垂线	艉垂標	after perpendicular, AP
艉灯	艉燈	stern light
艉封板	艉封板,艉橫板	stern transom plate
艉管	艉軸管	stern tube
艉管螺母	艉軸管環首螺帽	stern-tube nut
艉管密封装置	艉軸封	stern shaft sealing, stern shaft seal
艉管填料函	艉軸管填料函	stern-tube stuffing box
艉管轴承	艉軸套軸承	stern-tube bearing
艉横梁	艉梁	transom beam
艉滑道	艉斜道,艉坡道	stern ramp
艉滑轮	艉槽輪(布纜船)	stern sheave
艉机型船	艉機型船,艉機艙船	stern-engined ship, engined aft
艉尖舱	艉[尖]艙	after peak, after peak tank
艉尖舱舱壁	艉尖艙壁	after peak bulkhead
艉阱	艉凹艙(小艇)	cockpit
艉肋板	艉肋板	transom floor
艉楼	艉樓	poop
艉落	艉驟降	tipping
艉锚	艉錨	stern anchor
艉锚绞车	艉錨絞車	sternline winch, stern position winch
艉门		stern port
艉鳍	艉鳍	deadwood, skeg
艉翘	艉部上翹變形	cocking up of after body
艉倾	艉俯	trim by stern
艉斜浪	艉側浪	quartering sea
艉支架	艉托架(下水)	after poppet, aft poppet

祖国大陆名	台湾地区名	英文名
艉踵	艉踵(艉柱或舵柱底部)	aftfoot
艉轴(=螺旋桨轴)		
艉轴架	艉軸架,螺槳支架	propeller shaft bracket, propeller strut
艉轴架轴承	艉軸支架軸承	strut bearing
艉柱	艉柱	stern post
艉柱底骨	舵跟材	sole piece
卫生泵	衛生水泵	sanitary pump
卫生单元	衛生單元	sanitary unit
卫星导航	衛星航法	satellite navigation
未焊透	不完全穿透	incomplete penetration
未钎透	不完全穿透	incomplete penetration
紊流探测器	紊流探測器	turbulence detector
稳定板	穩定鰭	stabilizing fin, stabilizer fin
稳定回转阶段	穩定回旋階段	steady turning period
稳弧剂	電弧穩定劑	arc stabilizer
稳燃器	穩定器	stabilizer
稳态	穩態	steady state
稳心	定傾中心	metacenter
稳心半径	定傾半徑	metacentric radius
稳心曲线	定傾中心曲線圖	locus of metacenters, metacentric diagram
稳性	穩度	stability
稳性消失角	穩度消失角	angle of vanishing stability
涡轮增压器	渦輪增壓機	turbocharger
卧式内燃机	臥式引擎	horizontal engine
污泥柜	污油[泥]櫃	sludge tank
污水泵	污水泵	sewage water pump
污水舱	舭水艙	bilge water tank, bilge tank
污水井	舭水井	bilge well
污压载水	不潔壓艙水	dirty ballast
污液舱	污油[水]櫃	slop tank
钨极惰性气体保护焊	氣體遮蔽鎢弧焊	TIG welding
无杆锚	無桿錨,山字錨	stockless anchor
无缆系结装置	無纜系結裝置	ropeless linkage
无缆遥控潜水器	無纜遙控潛水器	untethered remotely operated vehicle
无人潜水器	無人潛水器	unmanned submersible
无人[值班]机舱	無人[當值]機艙空間	unattended machinery space
无塑性转变温度	無延性轉變溫度	nil-ductility transition temperature

祖国大陆名	台湾地区名	英 文 名
无损检验	非破壞檢查,非破壞試驗	nondestructive inspection, NDI, nondestructive testing, NDT
无线电导航	無線電導航	radio navigation
无线电信标	無線電示標	radio beacon
坞底	塢底	dock bottom, dock floor
坞墩	塢墩	docking block
坞坎	塢檻	dock sill
坞口		dock entrance
坞龙骨	駐塢龍骨	docking keel
坞门	塢門	dock gate, caisson
坞墙	浮塢牆	wing wall
坞室	塢室	dock chamber
坞首	塢首	dock head
坞修	塢修	docking repair, dock repair
误操作报警器	誤操作警報[器]	wrong operation alarm
雾化	霧化	atomization

X

祖国大陆名	台湾地区名	英 文 名
吸泥管	吸入管	suction pipe
吸气阀	進入閥	inlet valve
吸头	吸引水頭	suction head
吸扬挖泥船	吸管式挖泥船	suction dredger
牺牲阳极	陽極耗蝕	sacrificial anode
牺牲阳极利用效率	犧牲陽極利用效率	utilization coefficient for sacrificial anode
牺牲阳极阴极保护	犧牲陽極陰極保護	sacrificial anode cathodic protection, galvanic anode protection
舾装	舾裝	outfitting, hull fittings
舾装码头		fitting-out quay
舾装设备		outfit of deck and accommodation
舾装数	舾裝數,屬具數	equipment number
洗舱	洗艙	washing
洗炉	洗爐	washing boiler
系泊设备	系泊設備	mooring equipment
系泊设备试验	系泊裝置試驗	mooring arrangement test
系泊试验	系泊試車	mooring trial
系泊旋转接头	系船轉環	mooring swivel

祖 国 大 陆 名	台 湾 地 区 名	英 文 名
系泊状态	系泊狀態	moored condition
系船浮筒	系泊浮筒	mooring buoy
系船设备	錨泊及系泊設備	anchoring and mooring equipment
系缆具	系纜具	mooring fittings
系缆索	系泊索	mooring line
系索羊角	系索扣	cleat
系桩拉力	系纜拖力	bollard pull
狭航道效应	局限水域效應	restricted water effect
下导轮	下導輪	lower tumbler
下舵承	下舵承	neck bearing
下甲板	下甲板	lower decks
下降管	降水管(鍋爐)	downcomer, downcomer tube
下水	下水	launching
下水驳	下水駁船	launching barge
下水车	下水船架	launching cradle
下水重量	下水重量	launching weight
下死点	下死點	lower dead point
下止点	下死點	bottom dead center, BDC, lower dead center
纤维索	纖維纜線	fiber rope
弦长	弦	chord
弦杆	桁弦	chord
舷边动力滚柱	舷側動力滚子	side power roller
舷侧阀	舷側閥	ship side valve
舷侧肋骨	舷側縱材	side longitudinal
舷侧竖桁	側深橫肋	side transverse
舷侧外板	船側外板列	side plating
舷侧纵桁	側加強肋	side stringer
舷窗	舷[側圓]窗	side scuttle, side light
舷窗盖	舷窗內蓋,內窗蓋	deadlight
舷灯	舷燈	side light, side lamp
舷顶列板	舷側厚板列	sheer strake
舷弧	舷弧	sheer
舷门	舷側艙口	side port
舷门灯	梯口燈	gangway light
舷墙	舷牆	bulwark
舷墙门	梯口信道	gangway port
舷墙梯	舷牆梯	bulwark ladder

祖 国 大 陆 名	台 湾 地 区 名	英 文 名
舷伸甲板	張出甲板,舷外平台	sponson deck
舷梯	舷梯	accommodation ladder, gangway ladder
舷梯绞车	舷梯絞車	accommodation ladder winch
舷外挂机	舷外機	outboard engine
舷外跨距	伸出[舷外]距離	boom outreach, outreach
舷外排出阀	舷側排泄閥	overboard discharge valve
限界线	邊際線	margin line
限位开关	極限開關	limit switch
相对计程仪	相對計程儀	speed through the water log
相对旋转效率	[相]對轉[動]效率	relative rotative efficiency
箱形舱口盖	箱形艙蓋	pontoon hatchcover
箱形中桁材	箱形中線縱梁	duct center girder
镶套	襯套	bushing
响应谱	反應[波]譜	response spectrum
消磁船	消磁船	degaussing ship, degaussing vessel
消防泵	消防泵,滅火泵	fire pump
消防船	消防艇	firefighting ship, fire boat
消防控制室	火警控制室	fire-control room
消耗式冷剂冷藏集装箱	冷凍貨櫃,冷藏貨櫃	refrigeration container
消灭曲线	衰減曲線	curve of extinction
消声器	消音器	silencer
小水线面双体船	小水線面雙體船	small waterplane area twin hull, SWATH
小艇	小艇	small craft, cutter
斜舱壁	斜艙壁	sloping bulkhead
斜肋骨	斜肋骨	cant frame
斜梁	斜梁	cant beam
斜剖线	斜剖面線	diagonal, diagonal line
斜楔	尖劈,楔	wedge
斜置式内燃机	斜置引擎	inclined engine, diagonal engine
泄放旋塞阀	排泄旋塞	drain cock
泄漏	漏出	leak-off
泄漏电流	漏電流	leakage current
芯	芯,模芯	core
信号	信號	signal
信号绳	信號繩	signal line
信号桅	信號桅	signal mast
信号烟雾	煙[霧信]號	smoke signal
兴波阻力	興波阻力	wave making resistance

祖 国 大 陆 名	台 湾 地 区 名	英 文 名
兴波阻尼	興波阻尼	wave making damping
星形链	星形鏈	star chain
星形内燃机	星形引擎	radial engine
行李舱	行李間	luggage room
形变时效	應變老化	strain ageing
Z 形[操纵]试验	Z 形[操縱]試驗,蛇航 [操縱]試驗	zigzag test
T 形接头	T 形接頭,三通管接	T-joint
形状稳性力臂曲线	[穩度]交叉曲線	cross curves of stability
形状系数	形狀係數	form coefficient
型吃水	型吃水量,模吃水	molded draft
L 型传动	L 型傳動	L-drive
V 型传动	V 型驅動	V-drive
Z 型传动	Z 型驅動	Z-drive
A 型吊架	A 形構架(起重船)	"A" frame
L 型钢	角鋼	L-section, L-bar
T 型钢	T 型材	T-section, T-bar
X 型结点	X 型接頭	X-joint
Y 型结点	Y 型接頭	Y joint
型宽	型寬,模寬	molded breadth
型排水量	型排水量,模排水量	molded displacement
型排水体积	型排水體積	molded volume
U 型剖面	U 型剖面	U-section
型深	型深,模深	molded depth
型式	型,式	type
T 型天线	T 型天線	T-antenna, T-aerial
Z 型推进	Z 型推進	Z-peller propulsion
A 型显示	A 型顯示	A-scope
型线	型線,模線	molded lines
型线放样	放樣	laying-off of hull lines, laying down
型线光顺	線型整順	lines fairing
型线图	線型[圖]	lines plan, lines
型值	船線座標	offsets
型值表	船線座標表	table of offsets, offsets table
性能试验	性能試驗	performance test
汹涛阻力	洶濤阻力	rough-sea resistance, rough water resistance
修理船	修理艦	repair ship

祖国大陆名	台湾地区名	英 文 名
许可舱长	許可長度	permissible length
续航力	續航[能]力	endurance, cruising ability
蓄电池室	蓄電池室	battery room
悬臂自升式钻井平台	懸臂自升式鑽油平台	cantilever jack-up drilling unit, cantilever jack-up drilling platform
悬挂舵	懸舵,吊舵	spade rudder, hanging rudder
悬链锚腿系泊装置	懸垂法錨泊(鑽油台)	catenary anchor leg mooring, CALM
旋臂试验	[船模]強制回旋試驗	rotating-arm test
旋臂水池	旋臂水槽	rotating arm basin
旋风分离器	旋風分離器	cyclone separator
旋流器	旋流器,回旋式噴嘴	swirler
旋入管接头	外螺紋活管套節	male union
旋涡泵	旋流泵	regenerative pump, helical flow pump
旋涡阻尼	旋渦阻尼	eddy making damping
旋转变流机	旋轉式換流機	rotary convertor
旋转磁场	旋轉磁場	rotational magnetic field, rotating field
旋转视窗	旋轉視窗	clear-view screen
选择性腐蚀	選擇腐蝕	selective corrosion
熏舱	熏艙消毒	fumigation
巡航速度	巡航船速	cruising speed
巡洋舰	巡洋艦	cruiser
巡洋舰艉	巡洋艦型艉	cruiser stern
循环泵	循環泵	circulating pump
循环滑油舱(=循环滑油柜)		
循环滑油柜,循环滑油舱	循環[潤]滑油櫃	circulating lubrication oil tank
循环冷却系统	循環冷卻系統	circulation cooling system
循环水槽	環流水槽,回流水槽	circulating water channel, circulating tank, circulating water tank
训练船(=训练舰)		
训练舰,训练船	訓練船	training ship, training vessel

Y

祖国大陆名	台湾地区名	英 文 名
压焊	壓接	pressure welding
压痕	壓痕	indentation

祖 国 大 陆 名	台 湾 地 区 名	英 文 名
压力试验	壓力試驗	pressure test
压力箱	壓力水櫃	pressure tank
压力铸造	壓鑄,模鑄	die casting
压射冲头	柱塞	injection piston, plunger
压缩过程	壓縮過程	compression process
压缩行程	壓縮衝程	compression stroke
压缩压力	壓縮壓力	compression pressure
压载	壓載,壓艙	ballast
压载泵	壓載[水]泵,壓艙[水]泵	ballast pump
压载水舱	壓載艙	ballast water tank, ballast tank
压载水总管	壓載管路	ballast main line, ballast line
压载系统	壓艙水系統	ballasting system, ballast system
氩弧焊	氩[氣電弧]焊	argon-arc welding
烟道	煙道	gas pass, flue
烟管	煙管	smoke tube
烟管锅炉	火管鍋爐	fire-tube boiler
烟幕	煙幕	smokescreen
烟气分析	煙道氣分析	flue gas analysis
延迟裂纹	延遲龜裂	delayed crack, delayed cracking
沿海船	沿海船,沿岸航行船	coaster vessel, coaster, coasting service vessel
研磨	研光	lapping
盐水泵	鹽水泵,鹵水泵	brine pump
檐板	廉板	curtain plate
眼板	眼板	eye plate
眼环	環板	ring plate
扬弹机	彈藥升降機	ammunition hoist
阳极	陽極	anode
阳极保护	陽極防蝕	anodic protection
仰焊位置	仰姿	overhead position
氧化焰	氧化焰	oxidizing flame, oxiding flame
氧矛切割	氧吹管切割	oxygen lance cutting
氧气呼吸器	氧氣呼吸器	oxygen breathing apparatus
氧乙炔焊	氧乙炔焊接	oxy-acetylene welding
氧乙炔焰	氧乙炔焰	oxy-acetylene flame
样板	型板,樣板,模板	template
样条	壓條	batten

祖 国 大 陆 名	台 湾 地 区 名	英 文 名
样箱	模型打樣	mock-up
摇摆试验	橫搖試驗	rolling test, rolling experiment
遥控水雷	遙控水雷	remote-controlled mine
咬边	過熔低陷(焊接)	undercut
咬缸	缸膠著	cylinder sticking
药皮	包覆料	coating, coating material
药芯焊丝	含焊劑芯焊線,複合焊線	flux-cored wire, flux combined wire
业务衡准数	客船[艙區劃分]基準數	criterion of service numeral
叶背	葉背	back of blade
叶端	葉端	blade end
叶根	葉根	blade root
叶根空化	葉根空化	root cavitation
叶厚比	葉厚比	blade thickness ratio
叶轮	葉輪	bladed disc, bladed wheel
叶面比	葉面積比	blade area ratio
叶面参考线	葉片參考線	blade reference line
叶面空化	[螺槳]葉[正]面空化	face cavitation
叶片	葉片	blade
[叶片]出口角	葉片出流角	blade outlet angle
叶[片]高[度]	葉片高	blade height
[叶片]进口角	葉片入流角	blade inlet angle
叶频	葉頻	blade frequency
叶切面	葉片剖面	blade section
叶梢	葉尖	blade tip
叶梢空化	葉尖空化	tip cavitation
叶元体	葉片元素	blade element
曳纲滑轮	曳網滑輪	warp block
曳绳钓起线机	曳繩釣起繩機	trolling gurdy
曳绳钓渔船	曳繩釣漁船	trolling boat, troller
液舱	液艙	liquid tank
液浮陀螺仪	液體羅經	floated gyro, fluid compass
液化气体船	液化氣體船	liquefied gas carrier
液化石油气舱	液化石油氣艙	liquefied petroleum gas tank
液化石油气船	液化石油氣船	liquefied petroleum gas carrier, LPG carrier
液化天然气舱	液化天然氣艙	liquefied natural gas tank

祖 国 大 陆 名	台 湾 地 区 名	英 文 名
液化天然气船	液化天然氣船	liquefied natural gas carrier, LNG carrier
液货船	液貨船	tanker
液位计	液位計	level gauge
液压机	液壓機	hydraulic press, hydraulic machine
液压离合器	液壓離合器	hydraulically controlled clutch, hydraulic clutch
液压联轴器	液壓聯結器	oil injection coupling, hydraulic coupling
液压式调速器	液壓調速器	hydraulic governor
一次风	初級空氣	primary air
一次空气	主空氣	primary air
医院船	醫院船	hospital ship
移船锚	小錨	kedge anchor
移动绞机	移動絞機	shift winch
移动式辅燃气轮机	移動式輔燃氣［渦］輪機	mobile auxiliary gas turbine
移动式平台	移動式平台	mobile unit, mobile platform
移动式钻井平台	移動式鑽探平台,可動式鑽油台	mobile drilling unit, mobile drilling platform
乙炔发生器	乙炔發生器	acetylene generator
异常喷油	異常噴射	abnormal injection
逸流	逸流,泄漏	leakage
溢流管	溢流管	overflow pipe
阴极保护	陰極防蝕	cathodic protection
音频(=声频)		
音响信号器具	音響信號器具	sound signal instrument
引船驳岸	引船駁岸	ship-directional quay
引船绞盘	捲索絞盤	warping capstan
引航员软梯	引水［人］梯	pilot ladder
引航员升降装置	引水員升降機	pilot hoist
饮水舱	飲用水艙	drinking water tank
应急舱口	逃生艙口	escape hatch
应急操舵装置	應急操舵裝置,應急舵機	emergency steering gear
应急灯	應急燈	emergency light
应急电源	應急電源	emergency source of electrical power, emergency electrical power source
应急电站	應急電站	emergency electrical power plant
应急舵	應急舵	jury rudder

祖国大陆名	台湾地区名	英 文 名
应急发电机	應急發電機	emergency generator
应急辅助鼓风机	應急鼓風機	emergency standby blower, emergency blower
应急关闭系统	應急自動關閉裝置	emergency shutdown system, emergency automatic shut down device
应急航行机组		take-home engine unit
应急疏水试验	應急泄水試驗	emergency draining test
应急消防泵	應急消防泵	emergency fire pump
应力	應力	stress
应力腐蚀	應力腐蝕	stress corrosion
应力腐蚀断裂	應力腐蝕龜裂	stress corrosion cracking
应力集中	應力集中	stress concentration
荧光	熒光[性]	fluorescence
荧光屏	熒光屏	fluorescent screen, fluoroscope
硬钎焊	硬焊	brazing
硬化时间	硬化時間	curing time
永久磁铁	永久磁鐵	permanent magnet
涌浪补偿器	波浪補償器	swell compensator
优化	最適化	optimization
邮件舱	郵件艙	mail room
油船	油輪	oil tanker, oil carrier
油底壳	油池,油槽	oil sump
油动机	伺服馬達	servomotor
油封	油封	oil sealing
油冷	油冷	oil cooling
油密舱壁	油密艙壁	oil tight bulkhead
油漆间	油漆庫	paint store
油、散、矿船	礦砂散裝貨與油兼用船	ore/bulk/oil carrier, OBO
油水分离设备	油水分離裝置	oily-water separating equipment
油水界面探测器	油水界面計	oil-water interface detector
油污水分离装置	油水分離器	oily water separator
油箱	油櫃,油艙	oil tank
油性混合物	含油混合物	oily mixture
油压试验	驗油,油驗	oil test
油渣泵	污油泵	sludge pump
鱿鱼钓机	鱿魚釣機	squid angling machine
鱿鱼钓渔船	鱿[魚]釣船	squid angling boat
游步甲板	散步甲板	promenade deck

祖国大陆名	台湾地区名	英 文 名
游览船(=旅游船)		
游艇	遊艇	pleasure craft, yacht
有酬载荷	酬载	payload
有挡锚链	日字鏈	stud chain
有杆锚	有桿錨	stock anchor
有效波倾角	有效波斜度	effective wave slope
有效干舷	有效乾舷	effective freeboard
有效功率	有效功率	effective power
有效热效率	有效熱效率	effective thermal efficiency
有效射程	有效射程	effective range
有效推力	有效推力	effective thrust
有效载荷	有效負載	payload
有效周期	有義波周期	significant wave period
有义波高	有義波高	significant wave height
右焊法	反向焊接	backhand welding
右开式	右開式	right hand model
右舷	右舷	starboard
右旋	右旋	right-hand turning
余高	焊縫凸量	reinforcement
余热锅炉	廢熱鍋爐	waste heat boiler
鱼舱	魚艙	fish hold
鱼雷快艇	魚雷[快]艇	torpedo boat, torpedo motor boat
鱼探仪	魚[群]探[測]儀	fish finder
渔船	漁船	fishing vessel, fishing boat
渔港监督艇		fishing port supervision boat
渔业灯	捕魚燈	fishing light
渔业调查船	漁業研究船	fishery research vessel, fisheries research boat
渔业加工船	漁[獲]加工船,漁業加工船	fish factory ship, fisheries factory ship, fishery factory ship
渔业实习船	漁業訓練船	fishery training vessel, fisheries training boat
渔业指导船	漁業指導船	fishery guidance ship, fisheries guidance boat
预冷器	預冷器	pre-cooler
预燃[燃烧]室	預燃室	prechamber, pre-combustion chamber
预热	預熱	preheat, preheating
预热器	預熱器	preheater

祖国大陆名	台湾地区名	英文名
预热送风	熱風	hot blast
预舾装	預舾裝	pre-outfitting
原油洗舱	原油洗艙	crude oil washing, COW
原油洗舱系统	原油洗艙系統	crude oil cleaning system, crude oil washing system
原油油船	原油輪	crude oil tanker, crude oil carrier, dirty tanker
原子氢焊	氫原子焊	atomic-hydrogen arc welding, atomic hydrogen welding
圆舭艇型	圓舭型船體	round-bilge hull
圆筒仓	[散裝貨]儲艙	silo
圆锥离合器	錐形離合器	cone clutch
远距离控制	遙控	remote control
远洋船	遠洋船舶	ocean going ship, ocean going vessel
云状空化	雲狀空化	cloud cavitation
运动补偿设备	運動補償設備	motion compensation equipment
运煤船	運煤船	coal carrier, collier
运木船	木材運輸船, 運木船	timber carrier, log carrier
运输船	運輸船	transport ship
运畜船	牲口船	cattle carrier, cattle ship
运转试验	運轉試驗	running trial

Z

祖国大陆名	台湾地区名	英文名
杂货船	雜貨船	general cargo ship
杂散电流腐蚀	迷走電流腐蝕	stray-current corrosion
杂物吊杆	吊柱, 吊架	davit
载驳船	駁船搬運船, 子母船, 浮貨櫃船	barge carrier, lighter aboard ship, LASH
载荷曲线	①負荷曲線 ②載重曲線	load curve
载货甲板	載貨甲板	cargo deck
载货量	載貨量	cargo deadweight
载冷剂泵	冷卻劑泵(原子爐)	coolant pump
载人潜水器	有人潛水器	manned submersible
载体	載具	vehicle
载重量	載重[量]	deadweight

祖 国 大 陆 名	台 湾 地 区 名	英 文 名
载重量标尺	载重標尺	deadweight scale
载重线	载重[水]線	load line
载重线标志	载重線標誌	load line mark
再热	再熱	reheat
再热器	再熱器	reheater
再生式热交换器	再生式熱交換器	regenerative heat exchanger
凿削	鑿平	chipping
早燃	先期點火	pre-ignition
造波机	造波機	wave generator, wave maker
造型	造模翻砂,成型	molding
造型机	造模機,塑造機	molding machine
噪声	噪音	noise
[增]压比	壓力比	pressure ratio
增压泵	增壓泵	booster pump
增压锅炉	增壓鍋爐	supercharged boiler
增压内燃机	增壓引擎	supercharged engine
增压压力	增壓壓力	boost pressure
增益	增益	gain
轧辊	滾子,輥子	roller
轧制	軋	rolling
闸式潜水器	閘式潛水器	lock-in lock-out submersible
展开面积	展開面積	developed area
展弦比	展弦比	aspect ratio
战斗舰艇	軍艦,戰艦	combatant ship, combatant vessel
栈桥式码头	突堤碼頭	jetty
张紧器	張力器	tensioner
张力腿平台	張力腳式鑽油台	tension leg platform, TLP
罩壳	蓋	cover
着火	點火	ignition
遮蔽	遮蔽	masking
遮障	隔板,隔屏	screen
折边肘板	卷邊腋板,凸緣腋板	flanged bracket
折叠式舱口盖	折疊[式]艙口蓋	folding hatchcover
折角型艇	硬棱船體	hard-chine hull
针孔	針孔,銷孔	pinhole
侦察船	偵查艦	scout ship, scout
真空度	真空度	vacuum, degree of vacuum
真空冷凝器	真空冷凝器	vacuum condenser

祖 国 大 陆 名	台 湾 地 区 名	英 文 名
真空破坏器	真空破除器	vacuum breaker
真空试验	真空試驗	vacuum test
振动	振動	vibration
振动烈度	振動烈度	vibration severity
振动允许界限	振動容許限度	allowable limits of vibration
蒸馏器	蒸餾器	distiller
蒸馏水柜	蒸餾水櫃	distilled water tank, distilling tank
蒸汽动力装置	蒸汽動力裝置	steam power plant
蒸汽锅炉	蒸汽鍋爐	steam boiler
蒸汽机船	汽船,輪船	steam ship, steamer
蒸汽机动力装置	蒸汽機動力設備	steam engine power plant
蒸汽警笛	汽笛	steam whistle
蒸汽滤器	濾汽器	steam strainer
蒸汽灭火系统	蒸汽窒火系統	steam fire extinguishing system, steam smothering system
蒸汽室	汽櫃	steam chest
蒸－燃联合动力装置	組合蒸汽燃氣渦輪機動力設備	combined steam and gas turbine power plant
整流器	整流器	rectifier
整流罩舵	整流舵,球形舵	bulb-type rudder, counter rudder, bulb rudder
整体式浮船坞		single unit floating dock
正铲挖泥船	戽斗挖泥船	dipper dredger
正常状态	正常狀況	normal condition
正浮	縱平浮	floating on even keel, zero trim
正接	正極性	straight polarity
正面角焊缝	正面填角焊道	fillet weld in normal shear, front fillet weld
直接连接	直接連接	lug connection
直接数控	直接數值控制	direct numerical control, DNC
直径系数	直徑因數(螺槳)	Taylor's diameter constant, Taylor's advance coefficient, diameter factor
直立型艏	直立型艏	vertical bow
直列式内燃机	直列引擎,單排引擎	in-line engine
直流电动发电机	電動發電機	dynamotor
直流电动机	直流馬達	direct current motor
直流电力推进装置	直流電力推進裝置	D. C. electric propulsion plant
直流发电机	直流發電機	direct current generator

祖 国 大 陆 名	台 湾 地 区 名	英 文 名
直流扫力	單流驅氣	uniflow scavenging
直升机甲板	直升機甲板	helicopter deck
直梯	直立梯	vertical ladder
指挥室围壳	[潛艇]指揮室	sail
指挥塔台	指揮塔	command tower, direct tower
指令舵角	指令舵角	ordered rudder angle
指示功率	指示馬力	indicated power, indicated horsepower, IHP
指示热效率	指示熱效率	indicated thermal efficiency
指示油耗率	指示馬力燃油消耗率	indicated specific fuel consumption, ISFC
制荡板	制水板	swash plate
制荡舱壁	制水艙壁	swash bulkhead
制动力矩	制動扭矩	brake torque
制动装置试验	制動馬力試驗	brake test
制冷剂泵	冷媒泵	refrigerating medium pump
制流板	制水板	swash plate
质量控制	品[質]管[制]	quality control
中垂	舯垂[現象]	sagging
中拱	舯拱	hogging
中桁材	中線縱梁	center girder
中横剖面	舯[橫]剖面	midship section
中横剖面系数	舯剖面係數	midship section coefficient
中弧线	拱線	camber line
中机型船	舯機艙船	amidships-engined ship
中间肋骨	中間肋骨	intermediate frame
中间冷却器	中間冷卻器	intercooler
中间轴	中間軸	intermediate shaft
中间轴承	中間軸承	intermediate shaft bearing, plummer block
中内龙骨	中線內龍骨	center keelson, center line keelson
中线面	中線面	center line plane, center plane
中心系泊定位钻井船	中心系泊定位鑽探船	center moored drilling ship, turret moored drilling ship
中性焰	中性火焰	neutral flame
中压透平	中壓渦輪機	intermediate pressure turbine
中站面	中站面	midstation plane
中纵剖面	中心縱剖面	longitudinal section in center place
终端盒	接線盒	terminal box
舯剖面惯性矩	舯剖面慣性矩	moment of inertia of midship section

祖 国 大 陆 名	台 湾 地 区 名	英 文 名
重力焊	重力式[電弧]焊接	gravity welding
重力式吊艇架	重力[小艇]吊架	gravity-type davit, gravity type boat davit
重力式平台	重力式鑽油台	gravity platform
重力式下水	重力下水	gravity launching
重力油柜	重力油櫃	gravity oil tank
重量代换系统	重量代換系統	weight replacing system
重量吨	重量噸	weight ton
重量曲线	重量曲線	weight curve
重稳距	定傾[中心]高度	metacentric height
重心	重心	center of gravity
重心垂向坐标	垂向重心	vertical center of gravity
重心横向坐标	横向重心	transverse center of gravity
重心纵向坐标	縱向重心	longitudinal center of gravity
重要构件	主構件	primary member
洲际导弹	洲際彈道飛彈	intercontinental missile
轴承刮削	軸承刮削	scraping of bearing
轴承合金	軸承合金	bearing metal, antifrictional metal
轴承力	軸承力	bearing force
轴毂	螺[槳]轂	propeller boss
轴流式汽轮机	軸流渦輪機	axial flow turbine
轴隧	軸道	shaft tunnel
轴隧舱室	軸道凹部	tunnel recess
轴套	軸套, 軸襯	shaft liner
轴系	軸系	shafting
轴系传动装置	軸系傳動裝置	transmission gear of shafting
轴系校中	軸線校準, 軸線校中	centering for shafting, shafting alignment, shaft alignment
轴系接地装置	軸接地裝置	shafting-grounding device
轴系扭转振动	軸系扭轉振動	torsional vibration of shafting
轴系效率	軸系效率	transmission efficiency of shafting, shafting efficiency
轴系找中	軸系對準, 對中, 看中	centering of shafting
轴系振动	軸系振動	shafting vibration
轴系制动器	軸遊轉防止裝置	shafting brake, shaft-locking device
轴系纵向振动	軸系縱向振動	axial vibration of shafting, longitudinal vibration of shafting
轴线上叶厚	軸線上葉厚	blade thickness on axial line

祖 国 大 陆 名	台 湾 地 区 名	英 文 名
轴向推力	軸向推力	axial thrust
肘板	肘板,腋板	bracket, knee, knee plate
主操舵装置	主操舵裝置	main steering gear
主柴油机	主柴油機	main diesel engine
主尺度	主要尺寸	principal dimensions
主尺度比		dimension ratio
主船体	主船殼	main hull
主电源	主電源	main source of electrical power, main electric power source
主动舵	主動舵	active rudder
主舵杆	舵葉主構體	rudder main stock
主发电机组	主發電機組	main generating set
主干电缆	主電纜	main cable
主给水系统	主給水系統	main feed system
主钩起重量	主鈎起重量	main hook load
主锅炉	主鍋爐	main boiler
主机	主機	main engine
主机基座	主機座,主機台	main engine foundation, main engine bed
主机校中		determination for main engine location
主空气瓶	主空氣瓶	main air bottle
主肋骨	主肋骨	main frame
主令电器	主控開關,總開關	master switch
主令控制器	主控制器	master controller
主罗经	主羅經	master compass
主配电板	主配電盤	main switchboard
主汽阀	主停止閥	main stop valve
主汽轮机	主汽輪機	main steam turbine
主汽轮机组	主[蒸]汽[渦]輪機組	main steam turbine set
主燃气轮机	主燃氣渦輪機	main gas turbine
主燃气轮机组	主燃氣渦輪機組	main gas turbine set
主压载水舱	主壓載艙	main ballast tank
主要构件	主構件	primary member
主要要素	主要特徵,主要項目	principal particulars
注入管	注入管	filling piping, filling pipe
注水器	注射器	injector
注水式船坞		flooding dock
驻波	駐波	standing wave, stationary wave
柱塞泵	柱塞泵	plunger pump

祖 国 大 陆 名	台 湾 地 区 名	英 文 名
柱稳半潜式钻井平台	柱穩定型鑽油船	column stabilized semi-submersible drilling unit, column stabilized type drilling rig
铸钢	鑄鋼	cast steel
铸件	鑄件	casting, cast
铸件线收缩率	收縮量	shrinkage
铸铁	鑄鐵	cast iron
铸型	鑄模	mold
铸造	鑄造	foundry, casting
抓斗	抓斗	grab
抓斗挖泥船	抓斗[式]挖泥船	grab dredger, grapple dredger
抓斗稳索	抓索	grab stabilizer line, grab line
专用清洁压载泵	清潔壓艙水泵	clean ballast pump
专用压载水	隔離壓艙水	segregated ballast
专用压载水舱	隔離壓載水艙	segregated ballast tank, SBT
转出式吊艇架	旋臂吊架	radial davit
转动导缆孔	萬向導纜孔	universal chock
转动惯量	慣性矩	moment of inertia
转动套环	押環(汽旋機)	shroud ring
转矩系数	轉矩係數	torque coefficient
转矩仪	轉矩計	torque meter
转艏性	航向改變性	course changing quality
转速变换器	變速器	speed changer
转速记录器	轉速計	tachometer
转速调整特性	調速特性	speed regulation characteristic
转台	轉台	turntable
转柱舵	轉柱舵	rotating cylinder rudder
转子	轉子	rotor
转子式气笛	氣笛	air whistle
桩腿	樁腿	spud leg
装配	裝配	assembly
装载仪	負荷指示器	load indicator
坠落试验	墜落試驗	drop test
自动操舵仪	自動操舵裝置	autopilot
自动测向仪	自動測向儀	automatic direction finder, ADF
自动焊	自動焊接	automatic welding
自动控制	自動控制	automatic control
自动停车装置	自動停車裝置	automatic shut-down device

祖 国 大 陆 名	台 湾 地 区 名	英 文 名
自动系泊绞车	自動系泊絞車	automatic mooring winch
自航船(＝机动船)		
自航发射	［魚雷］自滑發射	swim-out discharge, swim-out
自航试验	自推試驗(船模試驗)	self-propulsion test
自航挖泥船	自航式挖泥船	self propelled dredger, self propelling dredger
自航因子	自推因子	self-propulsion factor
自亮浮灯	［救生圈］自燃燈	self-igniting buoy light
自流式循环水系统	自然式循環水系統	scoop circulating water system
自抛光防污涂料	自拋光防污漆,自磨型防污漆	self polishing copolymer antifouling paint
自起动装置	自動起動器	automatic starting device, automatic starter
自然通风	自然通風	natural draft, natural draught
自升式钻井平台	升降式鑽油台	jack-up drilling unit, jack-up rigs
自调	自動調節	self-regulation
自由活塞空气压缩机	自由活塞壓縮機	free piston air compressor, free piston compressor
自由液面	自由液面	free surface
自由液面修正	自由液面修正	free surface correction
自由自航船模试验	自航船模［運動］試驗	free-running model test
自重	空重,皮重	tare mass, tare weight
综合导航	整合航海	integrated navigation
综合调查船		comprehensive research vessel, comprehen sive research ship
总布置	總佈置［圖］	general arrangement
总长	全長	length overall, LOA
总段	船段結構,船［體分］段	block
总段建造法	［船體］分段建造法	block method of hull construction
总段舾装	船段舾裝	block outfitting
总吨位	總噸位	gross tonnage
总排水量	總排水量	total displacement
总用泵	通用泵,常用泵	general service pump
总振动		global vibration
总纵强度	縱向強度	longitudinal strength
总纵强度试验	縱向強度試驗	longitudinal strength test
总纵弯曲	縱［向］彎曲	longitudinal bending
总纵弯曲正应力	縱向彎應力	normal stress due to longitudinal bending

祖国大陆名	台湾地区名	英 文 名
		moment, longitudinal bending stress
总阻力	總阻力	total resistance
纵舱壁	縱[向]艙壁	longitudinal bulkhead
纵荡	縱移	surging
纵骨	縱材	longitudinal
纵骨架式	縱肋系統	longitudinal framing system
纵桁	縱桁	longitudinal girder
纵距	前進距離(回旋圈)	advance
纵剖线	縱剖面線	buttocks, buttock line
纵倾	俯仰	trim
纵倾角	俯仰角	angle of trim
纵倾力矩	俯仰力矩	trimming moment
纵倾平衡泵	俯仰水泵	trimming pump
纵稳心	縱定傾中心	longitudinal metacenter
纵稳性	縱向穩度	longitudinal stability
纵向滑道	縱向滑道	end slipway
纵向强力构件	縱向強度構件	longitudinal strength member
纵向下水	艉向下水	end launching
纵向振动	縱向振動	longitudinal vibration
纵斜	後傾(螺葉)	rake
纵摇	縱搖	pitching
纵重稳距	縱定傾中心高	longitudinal metacentric height
阻力峰	駝峰(速力曲線)	hump
阻力换算修正值	關聯裕度	correlation allowance model-ship, correlation allowance
阻力试验	阻力試驗	resistance test
阻力系数	阻力係數	resistance coefficient
阻塞	阻塞	choking, blockage
阻塞比	阻塞比	blockage ratio
阻塞效应	阻塞效應	blockage effect
阻塞修正	阻塞修正	blockage correction
阻索器	停止器	stopper
组合肋板	空架[底]肋板	bracket floor
组合起动屏	群起動盤	group starter panel
钻井驳	鑽油駁	drilling barge
钻井船	鑽油船	drilling ship, drilling vessel
钻井供应船	鑽油台補給船	drilling tender
钻井平台	鑽油平台,鑽油台	drilling platform, drilling unit, drilling rig

祖 国 大 陆 名	台 湾 地 区 名	英 文 名
钻探船	鑽油船	drilling ship, drilling vessel
钻削	鑽鑿	drilling
钻中心孔	定心	centering
最大长	全長	extreme length
最大持续功率	連續最大出力,額定最 大連續出力	maximum continuous rating, MCR
最大复原力臂角	最大扶正力臂角	angle of maximum righting lever
最大横剖面	最大橫剖面	maximum section
最大横剖面系数		maximum transverse section coefficient
最大宽	最大寬度,全寬	extreme breadth
最大叶宽	最大葉寬	maximum width of blade, maximum blade width
最大叶宽比	最大葉寬比	maximum blade width ratio
最高燃烧压力	最高燃燒壓力	maximum combustion pressure
最高转速	最高速[率]	maximum speed
最近陆地	最近陸地	nearest land
最深分舱载重线	最深艙區劃分載重線	deepest subdivision loadline
左焊法	前進焊法,[右手]左向 焊法	forehand welding
左开式	左開式	left hand model
左旋	左旋	left-hand turning
左转机组	左轉機組	left-hand revolving engine unit
作战能力	戰鬥性能	combat capability
作战情报中心室	戰情中心	combat information center, CIC
坐底式钻井平台	坐底式鑽油平台	submersible drilling unit, submersible dri- lling platform
坐坞强度	坐塢強度	docking strength
座板	底座	seat

副 篇

A

英　文　名	祖国大陆名	台湾地区名
abnormal injection	异常喷油	異常噴射
accelerated corrosion test	加速腐蚀试验	加速腐蝕試驗
accessibility	可达性	可達性
accommodation	起居舱室	起居艙[設施],住艙
accommodation deck	起居甲板	住艙甲板,起居艙甲板
accommodation equipment	舱室设备	
accommodation ladder	舷梯	舷梯
accommodation ladder winch	舷梯绞车	舷梯絞車
accommodation outfitting	内舾装	
accuracy control	精度管理	精度控制
acetylene generator	乙炔发生器	乙炔發生器
acidity	酸度	酸度
acid proof (= acid resistance)	防酸性能	耐酸,抗酸
acid resistance	防酸性能	耐酸,抗酸
acoustic frequency (= audio frequency)	声频,音频	聲頻
acoustic wave	声波	聲波,音波
active rudder	主动舵	主動舵
ACV (= air cushion vessel)	气垫船	氣墊船
adaptive control	适应控制	適應控制
added mass	附加质量	附加質量
ADF (= automatic direction finder)	自动测向仪	自動測向儀
adhesion	附着力,结合力	黏著[力],附著[力]
adjustable-speed motor	调速电动机	調速馬達
Admiralty coefficient	海军系数	海軍常數,海軍係數
advance	纵距	前進距離（回旋圈）
advance coefficient	进速系数	前進係數
advance ratio	进速比	前進比（螺槳）
aeroplane carrier (= aircraft carrier)	航空母舰	航空母艦

英　文　名	祖国大陆名	台湾地区名
afloat joining ship sections	水上合拢	
"A" frame	A型吊架	A形構架（起重船）
after body	后体	後半段船體
aftercooler	后冷却器	後冷卻器
after draft	艉吃水	艉吃水
after draught（＝after draft）	艉吃水	艉吃水
after mast	后桅	後桅
after peak	艉尖舱	艉［尖］艙
after peak bulkhead	艉尖舱舱壁	艉尖艙壁
after peak tank（＝after peak）	艉尖舱	艉［尖］艙
after perpendicular（AP）	艉垂线	艉垂標
after poppet	艉支架	艉托架（下水）
aftfoot	艉踵	艉踵（艉柱或舵柱底部）
aft poppet（＝after poppet）	艉支架	艉托架（下水）
ageing	老化	老化
agricultural vessel	农用船	
air charging test	充气试验	充氣試驗
air cooler	空气冷却器	空氣冷卻器
air cooling	空冷	氣冷
aircraft carrier	航空母舰	航空母艦
air cushion（＝cushion）	气垫	氣墊
air cushion vehicle（＝air cushion vessel）	气垫船	氣墊船
air cushion vessel（ACV）	气垫船	氣墊船
air duct	气道,风道	導氣管,通風管
air ejector	抽气器	空氣抽射器
air exhauster	空气抽除装置	抽氣器
air extractor（＝air exhauster）	空气抽除装置	抽氣器
air filter	空气滤器	空氣過濾器,空氣濾清器
air hammer（＝pneumatic hammer）	空气锤	［空］氣錘
air heater	①空气预热器 ②空气加热器	①空氣預熱器 ②空氣加熱器
air hole	透气孔	氣孔
air horn	气笛	汽笛,氣笛,號笛
air inlet unit	进气装置,进气设备	進氣裝置
air jet propelled boat	喷气推进船	噴氣推進艇
air lift mud pump	气力提升泵	吸泥泵

英 文 名	祖国大陆名	台湾地区名
air pipe	空气管	空氣管
air pipe head	空气管头	空氣管頭
air preheater（＝air heater）	①空气预热器 ②空气 加热器	①空氣預熱器 ②空氣 加熱器
air quenching	风冷	空氣淬火
air resistance	空气阻力	空氣阻力
air screw	空气螺旋桨	空氣螺槳
air tank	浮力舱	空艙
air test（＝air charging test）	充气试验	充氣試驗
air vent	排气槽	通氣孔,通氣口
air whistle	转子式气笛	氣笛
alarm unit	报警装置	警報裝置
alkalinity	碱度	鹼度
all-direction propeller	全回转推进器	全向螺槳
alleyway（＝passage）	通道	通道
allowable limits of vibration	振动允许界限	振動容許限度
all-round light	环照灯	環照燈
alternating current generator	交流发电机	交流發電機
alternating current motor	交流电动机	交流電動機
altiperiscope	对空潜望镜	對空潛望鏡
altitude above deck	甲板高度	
aluminum bronze	铝青铜	鋁青銅
ambient pressure	环境压力	環境壓力,周圍壓力
ambient temperature	环境温度	環境溫度,周圍溫度
amidships-engined ship	中机型船	舯機艙船
ammunition carrier room（＝ammunition handling room）	弹药转运间	彈藥搬運室
ammunition handling room	弹药转运间	彈藥搬運室
ammunition hoist	扬弹机	彈藥升降機
amphibious warfare ships and crafts	登陆战舰艇,两栖作战 舰艇	兩栖作戰艦艇
anchor	锚	錨
anchorage	锚地	錨［泊］池
anchor ball	锚球	錨球
anchor bed	锚床	錨床
anchor boat	起锚船	拋錨艇
anchor buoy	锚浮标	錨位浮標
anchor cable	锚索	錨索,錨纜

英 文 名	祖 国 大 陆 名	台 湾 地 区 名
anchor capstan	起锚绞盘	起錨絞盤
anchor chain	锚链	錨鏈
anchor crane (= anchor davit)	吊锚杆	吊錨桿
anchor davit	吊锚杆	吊錨桿
anchored condition	锚泊状态	錨泊
anchor fluke	锚爪	錨爪
anchor hawser	锚缆	
anchor holding power	锚抓力	錨抓著力
anchoring (= anchored condition)	锚泊状态	錨泊
anchoring and mooring equipment	系船设备	錨泊及系泊設備
anchoring equipment	锚泊设备	錨泊設備
anchoring trial	抛锚试验	錨試驗
anchor light	锚灯	錨泊燈
anchor moored positioning	锚泊定位	
anchor rack	锚架	
anchor recess	锚穴	嵌錨穴
anchor shackle	锚卸扣	錨接環
anchor shaft (= anchor shank)	锚干	錨幹
anchor shank	锚干	錨幹
anchor stock	锚横杆	錨桿
anchor stopper	掣锚器	止錨器
anchor test (= anchoring trial)	抛锚试验	錨試驗
angle of attack	锚爪袭角	攻角
angle of heel (= angle of list)	横倾角	橫傾角
angle of incidence	入射角	入射角,攻角
angle of list	横倾角	橫傾角
angle of maximum righting lever	最大复原力臂角	最大扶正力臂角
angle of trim	纵倾角	俯仰角
angle of vanishing stability	稳性消失角	穩度消失角
anisotropy	各向异性	異向性
annealing	退火	退火
annulus drag loss	环壁阻力损失	環周損失
anode	阳极	陽極
anodic protection	阳极保护	陽極防蝕
anodizing	电化学氧化	陽極處理,陽極防蝕
antenna	天线	天線
anti-collision light	避碰灯	
anti-corrosion paint	防锈涂料	防銹漆,防蝕漆

英　文　名	祖国大陆名	台湾地区名
anti-fouling	防污	防污
anti-fouling paint	防污涂料	防污漆
antifrictional metal（＝bearing metal）	轴承合金	軸承合金
anti-icing device	防冰设施	防冰設備
anti-icing equipment（＝anti-icing device）	防冰设施	防冰設備
antinode	波腹	波腹
anti-singing edge	抗谐鸣边	抗諧鳴邊（螺槳）
anti-skid deck paint	甲板防滑涂料	甲板防滑塗料
anti-slip deck paint（＝anti-skid deck paint）	甲板防滑涂料	甲板防滑塗料
AP（＝after perpendicular）	艉垂线	艉垂標
appendage resistance	附体阻力	附屬物阻力
appendages	附体	附屬物
applied voltage	外施电压	外施電壓
arc brazing	电弧钎焊	電弧硬焊
arc cutting	电弧切割	電弧切割
arch	拱	拱
arc spot welding	电弧点焊	電弧點焊
arc stabilizer	稳弧剂	電弧穩定劑
arctic vessel	极区船	
arc voltage	电弧电压	電弧電壓
arc welder（＝arc welding machine）	［电］弧焊机	電焊機
arc welding	［电］弧焊	電［弧］焊［接］
arc welding machine	［电］弧焊机	電焊機
area ratio	面积比	面積比
argon-arc welding	氩弧焊	氬［氣電弧］焊
articulated column	铰接柱,铰接塔	關節連接柱
artificial intelligence	人工智能	人工智慧
artillery	火炮	火炮
A-scope	A 型显示	A 型顯示
ash content	灰分	含灰量
aspect ratio	展弦比	展弦比
assembly	装配	裝配
assembly frame	假舱壁	
assembly jig	胎架	組合模架
assisted circulation boiler（＝forced circulation boiler）	强制循环锅炉	強制循環鍋爐

英 文 名	祖 国 大 陆 名	台 湾 地 区 名
astern condition	倒车工况	
astern gas turbine	倒车燃气轮机	倒車燃氣渦輪機
astern guardian valve（＝astern valve）	倒车阀	倒車[保]護閥
astern nozzle	倒车喷嘴	倒車噴嘴
astern trial	倒车试验	倒車試航
astern valve	倒车阀	倒車閥
atmospheric condenser	大气冷凝器	大氣冷凝器，大氣凝結器
atmospheric diving suit	单人常压潜水服	大氣壓潛水衣
atmospheric pressure	大气压	大氣壓力
atomic-hydrogen arc welding	原子氢焊	氫原子焊
atomic hydrogen welding（＝atomic-hydrogen arc welding）	原子氢焊	氫原子焊
atomization	雾化	霧化
attemperator	减温器	減溫器，调溫器
attenuation	衰减	衰減
audible signal	声响信号	音響信號，聽覺信號
audio frequency	声频	聲頻
automatic control	自动控制	自動控制
automatic control system for marine electric power plant	船舶电站自动控制装置	
automatic direction finder（ADF）	自动测向仪	自動測向儀
automatic mooring winch	自动系泊绞车	自動系泊絞車
automatic shut-down device	自动停车装置	自動停車裝置
automatic starter（＝automatic starting device）	自起动装置	自動起動器
automatic starting device	自起动装置	自動起動器
automatic starting installation for electrical motor driven auxiliaries	电动辅机自动起动装置	電動輔機自動起動裝置
automatic welding	自动焊	自動焊接
autopilot	自动操舵仪	自動操舵裝置
auxiliary boiler	辅锅炉	輔鍋爐，副鍋爐
auxiliary diesel engine	辅柴油机	輔柴油機
auxiliary feed line（＝auxiliary feed system）	辅给水系统	輔給水系統
auxiliary feed system	辅给水系统	輔給水系統
auxiliary gas turbine	辅燃气轮机	輔燃氣渦輪機
auxiliary machinery compartment	辅机舱	輔機艙

英 文 名	祖国大陆名	台湾地区名
auxiliary machinery room (= auxiliary machinery compartment)	辅机舱	輔機艙
auxiliary ship (= auxiliary ship and service craft)	辅助舰船	輔助艦
auxiliary ship and service craft	辅助舰船,军辅船	輔助艦
auxiliary steam turbine	辅汽轮机	輔蒸汽渦輪機
auxiliary steam turbine set	辅汽轮机组	
auxiliary steering gear	辅助操舵装置	輔[助]操舵装置
availability	可用性	可用性
average	海损	海損(保險)
awning	天幕	天遮
awning curtains	天幕帘	天遮簾[幕]
awning rope	天幕索	天遮索
axial flow turbine	轴流式汽轮机	軸流渦輪機
axial thrust	轴向推力	軸向推力
axial vibration of shafting	轴系纵向振动	軸系縱向振動
axis of weld	焊缝轴线	焊接線
azimuth(= azimuth angle)	方位角	方位角
azimuth angle	方位角	方位角

B

英 文 名	祖国大陆名	台湾地区名
Babbitt metal	巴氏合金	巴比合金,白合金
back cavitation	背空化	[葉]背空蝕
back chipping	清根	背縫鑿淨,背縫鑿平
back fire	①回火 ②逆火	①回火 ②逆火
backhand welding	右焊法	反向焊接
backing run	打底焊道	打底焊接
backing weld (= backing run)	打底焊道	打底焊接
back of blade	叶背	葉背
back pressure	背压	背壓,反壓
back pressure regulator	背压调节器	背壓調整器
back-pressure turbine	背压式汽轮机	背壓渦輪機
back-up breaker	后备断路器	備用斷路器
back weld	封底焊道	背焊道
balanced draft	平衡通风	平衡通風
balanced rudder	平衡舵	平衡舵

英 文 名	祖国大陆名	台湾地区名
bale cargo capacity	包装舱容	包裝貨容積
ball joint	球形接头	球接頭
ballast	压载	壓載,壓艙
ballasting system	压载系统	壓艙水系統
ballast line (= ballast main line)	压载水总管	壓載管路
ballast main line	压载水总管	壓載管路
ballast pump	压载泵	壓載[水]泵,壓艙[水]泵
ballast system (= ballasting system)	压载系统	壓艙水系統
ballast tank (= ballast water tank)	压载水舱	壓載艙
ballast water tank	压载水舱	壓載艙
band plate	带板	系板
bar	棒料	棒料
bare hull	裸船体	裸船殼
bare terminal	焊条夹持端	焊條裸端
barge	驳船	駁船
barge carrier	载驳船	駁船搬運船,子母船,浮貨櫃船
barge unloading dredger	吹泥船	
bar keel	方龙骨	條龍骨
base line (BL)	基线	基線,基準線
base metal	母材	母材
base plane	基面	基準面
basic section	基准分段	
bathyscaphe (= deep diving submersible)	深潜器	深潛器
batten	样条	壓條
battery charging and discharging panel	充放电板	充放電盤
battery room	蓄电池室	蓄電池室
BDC (= bottom dead center)	下止点	下死點
bead	焊道	焊珠
beam knee	梁肘板	梁肘板,梁腋板
beam sea	横浪	横浪,舷浪
beam trawler	桁拖渔船	側拖網漁船
bearing force	轴承力	軸承力
bearing metal	轴承合金	軸承合金
bed plate	机座	底板,座板
behind ship test of propeller	船后螺旋桨试验	[螺槳]船後試驗
behind test (= behind ship test of propel-	船后螺旋桨试验	[螺槳]船後試驗

英　文　名	祖国大陆名	台湾地区名
ler)		
bell	号钟	號鍾
belt conveyor	带式输送机	帶式運送機
bench test	台架试验	試台試驗
bending	弯曲	彎曲
berth	泊位	泊位
berth assembly	船台装配	
berth outfitting	船台舾装	
bevel angle	坡口面角度	斜角
bilge	舭	舭
bilge block	舭墩	舭邊墩
bilge bracket	舭肘板	舭腋板
bilge keel	舭龙骨	舭龍骨
bilge main（＝bilge main line）	舱底水总管	舭水總管
bilge main line	舱底水总管	舭水總管
bilge pump	舱底泵	舭[水]泵
bilge radius	舭部半径	舭曲半徑
bilge strake	舭列板	舭板列
bilge system	舱底水系统	舭水系統
bilge tank（＝bilge water tank）	污水舱	舭水艙
bilge water	舱底污水	舭水,艙底水
bilge water tank	污水舱	舭水艙
bilge well	污水井	舭水井
billet	坯料	坯料
bipod mast	人字桅	人字桅,雙腳桅
BL（＝base line）	基线	基線,基準線
black heart malleable cast iron	黑心可锻铸铁	黑心展性鑄鐵
black water	黑水	污水
blade	①桨叶 ②叶片	①[槳]葉 ②葉片
blade area ratio	叶面比	葉面積比
bladed disc	叶轮	葉輪
bladed wheel（＝bladed disc）	叶轮	葉輪
blade element	叶元体	葉片元素
blade end	叶端	葉端
blade frequency	叶频	葉頻
blade height	叶[片]高[度]	葉片高
blade inlet angle	[叶片]进口角	葉片入流角
blade outlet angle	[叶片]出口角	葉片出流角

英 文 名	祖 国 大 陆 名	台 湾 地 区 名
blade reference line	叶面参考线	葉片參考線
blade root	叶根	葉根
blade section	叶切面	葉片剖面
blade thickness on axial line	轴线上叶厚	軸線上葉厚
blade thickness ratio	叶厚比	葉厚比
blade tip	叶梢	葉尖
blade tip clearance	梢隙	葉尖間隙
blade tip thickness	梢厚	葉尖厚
blank	毛坯	毛坯
blasting	喷射处理	噴砂處理
bleed	抽气	抽氣
blind flange	盲板法兰	管口蓋板
blind hole (=non-penetrated hole)	盲孔	盲孔
blind zone (=dead zone)	盲区	盲區
block	总段	船段結構,船[體分]段
blockage (=choking)	阻塞	阻塞
blockage correction	阻塞修正	阻塞修正
blockage effect	阻塞效应	阻塞效應
blockage ratio	阻塞比	阻塞比
block coefficient	方形系数	方塊係數
blocking arrangement	摆墩	擺墩,塢墩佈置
block load	搁墩负荷	擱墩負載
block method of hull construction	总段建造法	[船體]分段建造法
block model (=ship model)	船模	船模
block outfitting	总段舾装	船段舾裝
blocks	墩	墩
block sequence welding	分段多层焊	間段焊接法
block with sand box	砂箱墩	砂箱墩,砂墩
blow-by	窜气	漏氣
blow down	排污	沖放,放水
blowdown valve	排污阀	底部吹洩閥,沖放閥
blowhole	气孔	氣孔
blow off valve	放气阀	吹洩閥
blow-off system	放气系统	吹洩系統
blue shortness	蓝脆	藍脆性
boarding ladder	登艇梯	登艇梯
boat	艇	小艇,船艇
boat chock	艇座	小艇座

英　文　名	祖国大陆名	台湾地区名
boat davit	吊艇架	小艇吊架
boat deck	艇甲板	小艇甲板
boat deck lamp（＝boat deck light）	登艇灯	［艇筏］乘載照明燈,小艇甲板燈
boat deck light	登艇灯	［艇筏］乘載照明燈,小艇甲板燈
boat fall	吊艇索	小艇吊索
boat handing gear	吊艇装置	小艇吊放裝置
boat rope	固艇索具	艇艏系纜
boatswain's chair	吊板	工作吊板,單人吊板
boat winch	起艇绞车	小艇絞車
body lines	横剖线	
boiler	锅炉	鍋爐
boiler automatic control system	锅炉自动控制装置	鍋爐自動控制系統
boiler bearer（＝boiler foundation）	锅炉座	鍋爐座
boiler body（＝boiler proper）	锅炉本体	鍋爐本體
boiler efficiency	锅炉效率	鍋爐效率
boiler feed pump	锅炉给水泵	鍋爐給水泵
boiler forced circulating pump（＝boiler water forced circulating pump）	锅炉水强制循环泵	鍋爐強制循環泵
boiler foundation	锅炉座	鍋爐座
boiler fuel oil pump（＝fuel oil burning pump）	锅炉燃油泵	鍋爐燃油泵
boiler fuel oil system	锅炉燃油系统	鍋爐燃油系統
boiler proper	锅炉本体	鍋爐本體
boiler room	锅炉舱	鍋爐艙,鍋爐間
boiler room casing	炉舱棚	鍋爐艙棚,鍋爐艙圍壁
boiler saddle	锅炉支座	鍋爐鞍座
boiler water	锅水	［鍋］爐水
boiler water forced circulating pump	锅炉水强制循环泵	鍋爐強制循環泵
boiling out	［碱］煮炉	煮煉（除去爐中油脂雜質）
bollard	带缆桩	系纜樁
bollard pull	系桩拉力	系纜拖力
Bonjean's curves	邦戎曲线	龐琴曲線
boom outreach	舷外跨距	伸出［舷外］距離
booster gas turbine	加速燃气轮机	助力燃氣渦輪機
booster pump	增压泵	增壓泵

英 文 名	祖国大陆名	台 湾 地 区 名
boost pressure	增压压力	增壓壓力
boottopping corrosion (= waterline zone corrosion)	水线区腐蚀	水線區腐蝕
boottopping paint	水线涂料	水線漆
boring	镗孔	搪孔
boss (= hub)	桨毂	[螺槳]毂
bottom dead center (BDC)	下止点	下死點
bottom frame	船底横骨	[船]底肋骨
bottom longitudinal	船底纵骨	[船]底縱肋
bottom plating	船底板	[鋼]船底外板
bottom side tank	底边舱	
bottom transverse	船底横桁	船底横材
bow anchor	艏锚	艏錨,大錨,主錨
bow construction (= stem structure)	艏部结构	艏結構
bow door	艏门	
bow position winch	艏锚绞车	
bow rudder	艏舵	艏舵
bow sea	艏斜浪	艏側浪
bracing	撑杆	撑桿
bracket	肘板	肘板,腋板
bracket floor	组合肋板	空架[底]肋板
brake mean effective pressure	平均有效压力	制動平均有效壓力
brake test	制动装置试验	制動馬力試驗
brake torque	制动力矩	制動扭矩
brazing	钎焊,硬钎焊	硬焊
breadth depth ratio	宽深比	寬深比
breaking current	分断电流	切斷電流
breakwater	挡浪板	擋浪板
breather valve	呼吸阀	呼吸閥
bridge	桥楼	橋樓,駕駛台
bridge deck (= navigation deck)	驾驶甲板	航海甲板,橋樓甲板
bridge gauge	桥规	橋[形]規
brine pump	盐水泵	鹽水泵,鹵水泵
brittle fracture	脆性断裂	脆性破裂
broaching	①横甩 ②拉削	①横甩,横轉 ②拉削
brow (= eye brow)	眉毛板	窗楣
brush painting	刷涂	毛刷塗裝
bucket	泥斗	斗

英　文　名	祖国大陆名	台湾地区名
bucket dredger	链斗挖泥船	戽斗挖泥船
buffer	缓冲器	缓衝器
buffing	抛光	擦光
building berth	船台	造船台
build up welding（=surfacing）	堆焊	堆焊,堆焊接
built up sequence	熔敷顺序	组合顺序（焊接）
bulbous bow	球鼻艏	球[形]艏
bulbous stern	球形艉	球形艉
bulb rudder（=bulb-type rudder）	整流罩舵	整流舵,球形舵
bulb-type rudder	整流罩舵	整流舵,球形舵
bulk cargo	散装货物	散装貨
bulk cargo ship（=bulk carrier）	散货船	散装貨船
bulk carrier	散货船	散装貨船
bulkhead	舱壁	艙壁
bulkhead deck	舱壁甲板	艙壁甲板
bulkhead door	舱壁门	艙壁門
bulkhead plate	舱壁板	艙壁板
bulkhead recess	舱壁龛	艙壁凹入部
bulkhead stiffener	舱壁扶强材	艙壁防撬材
bulkhead stuffing box	隔舱填料函	艙壁填料函
bull trawler	双拖渔船	雙拖[網]漁船
bulwark	舷墙	舷牆
bulwark ladder	舷墙梯	舷牆梯
buoyancy	浮性	浮力
buoyancy curve	浮力曲线	浮力曲線
buoyancy force	浮力	浮力
buoyancy tank	浮箱	浮力艙櫃,浮箱
buoyant rescue quoit	施救浮索	營救浮環
buoyant smoke signal	漂浮烟雾信号	浮[式發]煙信號
buoy hook（=buoy shackle）	浮筒卸扣	浮筒系鈎
buoy shackle	浮筒卸扣	浮筒系鈎
buoy tender	航标船	浮標勤務船
burner	燃烧器	燃燒器
burner port	燃烧器喷口	燃燒器嘴
burnertip tip（=burner port）	燃烧器喷口	燃燒器嘴
burr	毛刺	毛口
busbar	汇流排	匯流排
bushing	镶套	襯套

英 文 名	祖 国 大 陆 名	台 湾 地 区 名
buttering	隔离堆焊层	預堆邊焊
buttock line（＝buttocks）	纵剖线	縱剖面線
buttocks	纵剖线	縱剖面線
butt welding	对接焊	對接焊接
by-pass control	旁通调节	旁通調節
by-pass governing（＝by-pass control）	旁通调节	旁通調節
by-pass valve	旁通阀	旁通閥

C

英 文 名	祖 国 大 陆 名	台 湾 地 区 名
cabin	居住舱室	房艙,艙室
cabin fan	舱室通风机	艙室[通]風機
cabin outfit	舱室属具	
cable box（＝cable expansion box）	电缆伸缩箱	電纜箱
cable coaming	电缆框,电缆筒	穿線環圍,線孔線圍
cable expansion box	电缆伸缩箱	電纜箱
cable layer	布缆船	布纜船
cable lifter	锚链轮	鏈輪
cable releaser	弃链器	釋纜器
cable ship（＝cable layer）	布缆船	布纜船
cable tank	电缆舱	電纜艙
CAD（＝computer-aided design）	计算机辅助设计	電腦輔助設計
cage mast	桁架桅	籠形桅
caisson	沉箱	沉箱,潛水箱
caisson（＝dock gate）	坞门	塢門
calibration	校准	校準
CALM（＝catenary anchor leg mooring）	悬链锚腿系泊装置	懸垂法錨泊（鑽油台）
CAM（＝computer aided manufacturing）	计算机辅助制造	電腦輔助製造
camber	梁拱	拱高
camber line	中弧线	拱線
cam clearance	凸轮间隙	凸輪間隙
cam shaft	凸轮轴	凸輪軸
cant beam	斜梁	斜梁
cant frame	斜肋骨	斜肋骨
cantilever jack-up drilling platform（＝cantilever jack-up drilling unit）	悬臂自升式钻井平台	懸臂自升式鑽油平台
cantilever jack-up drilling unit	悬臂自升式钻井平台	懸臂自升式鑽油平台

英　文　名	祖国大陆名	台湾地区名
can-type chamber（＝can-type combus-tor）	管形燃烧室	筒形燃燒室
can-type combustor	管形燃烧室	筒形燃燒室
capacity curve	容积曲线	容積曲線
capacity plan	舱容图	容積圖
CAPP（＝computer-aided process plan-ning）	计算机辅助工艺规程编制	電腦輔助制程規劃
capstan	绞盘	絞盤
captain room	船长室	船長室
captain's room（＝captain room）	船长室	船長室
captive model test	拘束船模试验	拘束模型試驗
carbon arc welding	碳弧焊	碳[極電]弧熔焊
carbon dioxide fire extinguisher	二氧化碳灭火器	二氧化碳滅火器
carbon fiber reinforced plastics	碳纤维增强塑料	碳纖維強化塑膠
carburization（＝carburizing）	渗碳	滲碳法
carburizing	渗碳	滲碳法
car deck（＝wagon deck）	车辆甲板	車輛甲板
cargo boat（＝cargo ship）	货船	貨船,貨輪
cargo capacity	货舱容积	貨艙容量
cargo carrier（＝cargo ship）	货船	貨船,貨輪
cargo deadweight	载货量	載貨量
cargo deck	载货甲板	載貨甲板
cargo hatch	货舱口	貨艙口
cargo hold	货舱	貨艙
cargo lamp（＝cargo light）	货舱工作灯	[裝卸]貨[照明]燈
cargo lift（＝cargo lift equipment）	起货设备	貨物升降機
cargo lift equipment	起货设备	貨物升降機
cargo light	货舱工作灯	[裝卸]貨[照明]燈
cargo oil handing system	货油装卸系统	貨油裝卸系統
cargo oil heating system（＝cargo oil tank heating system）	货油舱加热系统	貨油加熱系統
cargo oil hose	货油软管	貨油軟管
cargo oil main line	货油装卸总管	貨油總管
cargo oil pump	货油泵	貨油泵
cargo oil pumping system（＝cargo oil handing system）	货油装卸系统	貨油裝卸系統
cargo oil tank	货油舱	貨油艙
cargo oil tank cleaning installation（＝car-	货油舱洗舱系统	貨油艙洗艙裝置

英 文 名	祖 国 大 陆 名	台 湾 地 区 名
go oil tank cleaning system)		
cargo oil tank cleaning system	货油舱洗舱系统	貨油艙洗艙裝置
cargo oil tank gas-freeing installation （＝gas freeing）	驱气系统	貨油艙清除油氣裝置
cargo oil tank heating system	货油舱加热系统	貨油加熱系統
cargo oil tank stripping system	货油舱扫舱系统	貨油艙收艙系統
cargo oil tank venting piping system	货油舱透气系统	貨油艙通氣系統
cargo oil tank venting system（＝cargo oil tank venting piping system）	货油舱透气系统	貨油艙通氣系統
cargo oil transfer main pipe line（＝cargo oil main line）	货油装卸总管	貨油總管
cargo oil valve	货油阀	貨油閥
cargo purchase rigging	吊货索具	吊貨索具
cargo runner	吊货索	吊貨索
cargo ship	货船	貨船,貨輪
Cargo Ship Safety Certificate	货船安全证书	貨船安全證書
Cargo Ship Safety Construction Certificate	货船构造安全证书	貨船安全構造證書
Cargo Ship Safety Equipment Certificate	货船设备安全证书	貨船安全設備證書
Cargo Ship Safety Radio Certificate	货船无线电安全证书	貨船安全無線電證書
cargo sling（＝cargo runner）	吊货索	吊貨索
cargo space（＝cargo hold）	货舱	貨艙
cargo vessel（＝cargo ship）	货船	貨船,貨輪
cargo winch	起货绞车	吊貨絞機
cargo wire runner（＝cargo purchase rig- ging）	吊货索具	吊貨索具
carpenter store	木工间	木工庫
carriage	炮架	炮架
carrying capacity of lifeboat	救生艇乘员定额	救生艇容載量
cascade tank	热井	熱井,串列柜
case（＝enclosure）	外壳	外殼
cast（＝casting）	铸件	鑄件
casting	铸件	鑄件
casting（＝foundry）	铸造	鑄造
cast iron	铸铁	鑄鐵
castle（＝deck house）	甲板室	甲板室
cast steel	铸钢	鑄鋼
catamaran	双体船	雙［胴］體船
cat-chain（＝cat tackle）	吊锚索具	吊錨索具

英　文　名	祖国大陆名	台湾地区名
cat davit（=anchor davit）	吊锚杆	吊錨桿
catenary anchor leg mooring（CALM）	悬链锚腿系泊装置	懸垂法錨泊（鑽油台）
cat head（=anchor davit）	吊锚杆	吊錨桿
cathodic protection	阴极保护	陰極防蝕
cat tackle	吊锚索具	吊錨滑車組
cattle carrier	运畜船	牲口船
cattle ship（=cattle carrier）	运畜船	牲口船
cat walk（=connecting bridge）	天桥	天橋,連橋
caustic embrittlement	碱脆	鹼性脆化
cavitating propeller	空化螺旋桨	空化螺槳
cavitation	空化	空化
cavitation erosion	空蚀	空[泡沖]蝕
cavitation number	空化数	空化數
cavity	空泡	氣泡
celestial navigation	天文导航	天文導航
center girder	中桁材	中線縱梁
centering	校中	對中線
centering for shafting	轴系校中	軸線校準,軸線校中
centering of shafting	轴系找中	軸系對準,對中,看中
center keelson	中内龙骨	中線內龍骨
center line keelson（=center keelson）	中内龙骨	中線內龍骨
center line plane	中线面	中線面
center moored drilling ship	中心系泊定位钻井船	中心系泊定位鑽探船
center of buoyancy	浮心	浮[力中]心
center of floatation	漂心	浮面中心
center of gravity	重心	重心
center of rudder force（=center of rudder pressure）	舵压力中心	舵力中心
center of rudder pressure	舵压力中心	舵力中心
center of turning circle	回转中心	回轉中心
center plane（=center line plane）	中线面	中線面
centralized control	集中控制	集中控制
centralized control console of engine room	机舱集控台	
centrifugal casting（=centrifugal pressure casting）	离心浇注	離心鑄造法
centrifugal clutch	离心离合器	離心離合器
centrifugal compressor	离心[式]压气机	離心壓縮機
centrifugal pressure casting	离心浇注	離心鑄造法

英 文 名	祖国大陆名	台湾地区名
centrifugal pump	离心泵	離心泵
Certificate of Ship's Nationality	船舶国籍证书	船舶國籍證書
chain conveyor	链式输送机	鏈條輸送機
chain intermittent fillet weld	并列断续角焊缝	並列斷續填角焊接
chain locker	锚链舱	錨鏈艙
chain pipe	锚链管	錨鏈管,錨鏈筒
chain stopper	掣链器	錨鏈扣
chamot (= fireclay)	耐火黏土	耐火泥,火泥,燒磨土
chamotte (= fireclay)	耐火黏土	耐火泥,火泥,燒磨土
chamotte brick (= firebrick)	耐火砖	耐火磚,火磚,燒磨土磚
channel ship	海峡[渡]船	海峽船
channel steamer (= channel ship)	海峡[渡]船	海峽船
characteristic	特性	特性
characteristic impedance	特性阻抗	特性阻抗
charging and discharging board (= battery charging and discharging panel)	充放电板	充放電盤
chart house (= chart room)	海图室	海圖室
chart lamp (= chart table light)	海图灯	海圖燈
chart room	海图室	海圖室
chart table	海图桌	海圖桌
chart table light	海图灯	海圖燈
chemical cargo ship (= chemical tanker)	化学品船	化學品船
chemical carrier (= chemical tanker)	化学品船	化學品船
chemical cleaning	化学清洗	化學清洗
chemical fire extinguisher (= chemical reaction fire extinguisher)	化学反应式灭火器	化學滅火器
chemical reaction fire extinguisher	化学反应式灭火器	化學滅火器
chemical tanker	化学品船	化學液體船
chipping	①凿削 ②清铲	鑿平
chipping hammer	敲渣锤	鑿錘
chock	导缆钳	導[纜]索器
choking	阻塞	阻塞
chord	①弦长 ②弦杆	①弦 ②桁弦
chute	溜泥槽	滑槽
CIC (= combat information center)	作战情报中心室	戰情中心
CIMS (= computer integrated manufacturing system)	计算机集成制造系统	電腦整合製造系統
circulating lubrication oil tank	循环滑油柜,循环滑油	循環[潤]滑油櫃

英　文　名	祖国大陆名	台湾地区名
	舱	
circulating pump	循环泵	循環泵
circulating tank (＝circulating water channel)	循环水槽	環流水槽,回流水槽
circulating water channel	循环水槽	環流水槽,回流水槽
circulating water tank (＝circulating water channel)	循环水槽	環流水槽,回流水槽
circulation cooling system	循环冷却系统	循環冷卻系統
cladding (＝covering)	包覆	包覆,被覆
clad steel	复合钢	護面鋼
classification	入级	入級,船級
classification certificate	入级证书	船級證書
classification society	船级社	船級協會,船級機構
classification survey	入级检验	船級檢驗
class of ship	船级	船級
clean ballast	清洁压载水	清潔壓艙水
clean ballast pump	专用清洁压载泵	清潔壓艙水泵
clean ballast tank operation manual	清洁压载舱操作手册	清潔壓艙水操作手冊
clear-view screen	旋转视窗	旋轉視窗
cleat	系索羊角	系索扣
clip connection	面板切斜连接	短角鐵連接
close appliance (＝hull closures)	关闭设备	關閉裝置
cloud cavitation	云状空化	雲狀空化
clutch	离合器	離合器
coagulation	凝聚	凝聚
coal burning boiler (＝coal firing boiler)	燃煤锅炉	燃煤鍋爐
coal carrier	运煤船	運煤船
coal firing boiler	燃煤锅炉	燃煤鍋爐
coaster (＝coaster vessel)	沿海船	沿海船
coaster vessel	沿海船	沿海船
coasting service vessel (＝coaster vessel)	沿海船	沿岸航行船
coating (＝painting)	①药皮 ②涂装,涂料施工	①包覆料 ②塗裝
coating material (＝coating)	药皮	包覆料
coaxial cable	同轴电缆	同軸電纜
cocking up of after body	艉翘	艉部上翹變形
cocking up of forebody	艏翘	艏部上翹變形
cockpit	艉阱	艉凹艙（小艇）

英　文　名	祖国大陆名	台湾地区名
CODAG, CODOG (＝combined diesel and/or gas turbine power plant)	柴－燃联合动力装置	柴油燃氣渦輪組合機
code flag (＝signal flag)	号旗	信號旗
coefficient of balance of rudder	舵平衡比	舵平衡面積比
coefficients of form	船型系数	線型係數
cofferdam	空隔舱	堰艙
COGAG, COGOG (＝combined gas turbine and/or gas turbine power plant)	全燃联合动力装置	組合燃氣渦輪機
cohesion (＝coagulation)	凝聚	凝聚
cold brittleness	冷脆	冷脆性
cold cracking	冷裂	冷破裂
cold starting	冷态起动	常溫起動,冷車起動
cold storage boat (＝refrigerator ship)	冷藏船	冷藏船,冷凍船
cold store (＝refrigerating chamber)	冷藏库	冷藏庫,冷凍庫
cold work	冷作	冷作
collier (＝coal carrier)	运煤船	運煤船
collision bulkhead	防撞舱壁	防碰艙壁
COLREGS (＝International Regulations for Preventing Collisions at Sea)	国际海上避碰规则	國際海上避碰規則
column	①机柱 ②立柱	柱
column stabilized semi-submersible drilling unit	柱稳半潜式钻井平台	柱穩定型鑽油船
column stabilized type drilling rig (＝column stabilized semi-submersible drilling unit)	柱稳半潜式钻井平台	柱穩定型鑽油船
combatant ship	战斗舰艇	軍艦,戰艦
combatant vessel (＝combatant ship)	战斗舰艇	軍艦,戰艦
combat capability	作战能力	戰鬥性能
combat information center (CIC)	作战情报中心室	戰情中心
combined diesel and gas/or turbine power plant (CODAG, CODOG)	柴－燃联合动力装置	柴油燃氣渦輪組合機
combined gas turbine and/or gas turbine power plant (COGAG, COGOG)	全燃联合动力装置	組合燃氣渦輪機
combined power plant	联合动力装置	複合式動力裝置
combined steam and gas turbine power plant	蒸－燃联合动力装置	組合蒸汽燃氣渦輪機動力設備
combustion chamber	燃烧室	燃燒室
combustion efficiency	燃烧效率	燃燒效率

英　文　名	祖国大陆名	台湾地区名
combustion rate	燃烧率	燃燒率
combustor （ =burner）	燃烧器	燃燒器
combustor outer casing	燃烧室外壳	燃燒室外殼
command tower	指挥塔台	指揮塔
commercial ship （ =merchant ship）	商船	商船
compartment	舱	艙
compass	罗经,罗盘	羅經
compass bearing	罗方向	羅經方位［角］
compass course	罗航向	羅經航向
compass deck	罗经甲板	羅經甲板
compass repeater	分罗经	子羅經,羅經複示儀
compensating winding （ =compensating wire）	补偿电线	補償線圈
compensating wire	补偿电线	補償線圈
compensator	补偿器	補償器
complete annealing （ =full annealing）	完全退火	完全退火
composite oil-exhaust gas fired boiler	燃油废气组合式锅炉	廢氣燃油複合式鍋爐
composite oil-fired exhaust gas boiler （ =composite oil-exhaust gas fired boiler）	燃油废气组合式锅炉	廢氣燃油複合式鍋爐
comprehensive research ship （ =comprehensive research vessel）	综合调查船	
comprehensive research vessel	综合调查船	
compression pressure	压缩压力	壓縮壓力
compression process	压缩过程	壓縮過程
compression ratio	［几何］压缩比	壓縮比
compression stroke	压缩行程	壓縮衝程
computer-aided design （CAD）	计算机辅助设计	電腦輔助設計
computer aided manufacturing （CAM）	计算机辅助制造	電腦輔助製造
computer-aided process planning （CAPP）	计算机辅助工艺规程编制	電腦輔助制程規劃
computer controlled planar motion carriage	数控平面运动拖车	數控平面運動拖車
computer integrated manufacturing system （CIMS）	计算机集成制造系统	電腦整合製造系統
computerized planar motion carriage （ =computer controlled planar motion carriage）	数控平面运动拖车	數控平面運動拖車

英 文 名	祖国大陆名	台湾地区名
concrete gravity platform	混凝土重力式平台	混凝土重力式鑽油台
condensate pump	凝水泵	凝水泵
condenser	凝汽器	冷凝器,凝結器
condenser cooling surface	[凝汽器]冷却面积	冷卻面
condenser tube	冷却[水]管	冷凝管
condensing turbine	凝汽式汽轮机	凝汽渦輪機
conductivity	电导率	傳導性,傳導度
conductor	导体	導體
cone clutch	圆锥离合器	錐形離合器
connecting bridge	天桥	天橋,連橋
connecting rod	连杆	連桿
constant pitch	等螺距	定螺距
constant pressure cycle	定压循环	等壓循環
constant volume cycle	定容循环	等容循環
contactor	接触器	接觸器
container hold	集装箱舱	貨櫃艙
container lifting spreader（＝spreader）	集装箱吊具	貨櫃吊具
container ship	集装箱船	貨櫃船
container spreader（＝spreader）	集装箱吊具	貨櫃吊具
continuous casting	连续铸造	連續鑄造
continuous current（＝permanent current）	连续电流	連續電流
continuous member	连续构件	連續構件
continuous service rating（CSR）	持续常用功率	額定連續常用出力
continuous wave	连续波	連續波
continuous weld	连续焊缝	連續焊接
contrarotating propellers	对转螺旋桨	對轉螺槳
controllable-pitch propeller	可控螺距螺旋桨	[可]控[螺]距螺槳, [可]變螺距螺槳
controller	控制器	控制器
control relay	控制继电器	控制繼電器,控制電驛
control station	控制站	控制站,控制室
control system	控制系统	控制系統
conventional submarine	常规潜艇	傳統潛艇
convertor	变流机	換流機
coolant pump	载冷剂泵	冷卻劑泵（原子爐）
cooled blade	冷却叶片	冷卻葉片
cooling blade（＝cooled blade）	冷却叶片	冷卻葉片
cooling pipe（＝condenser tube）	冷却[水]管	冷卻管

英　文　名	祖国大陆名	台湾地区名
cooling rate	冷却倍率	冷卻[速]率
cooling surface (= condenser cooling surface)	[凝汽器]冷却面积	冷卻面
cooling water pump	冷却水泵	冷卻水泵
core	芯	芯,模芯
correlation allowance (= correlation allowance model-ship)	阻力换算修正值	關聯裕度
correlation allowance model-ship	阻力换算修正值	關聯裕度
corrosion	腐蚀	腐蝕,銹蝕
corrosion allowance	腐蚀裕量	腐蝕裕度
corrosion fatigue	腐蚀疲劳	腐蝕疲勞,銹蝕疲勞
corrosion prevention (= corrosion protection)	防蚀	防蝕
corrosion product	腐蚀产物	腐蝕生成物
corrosion protection	防蚀	防蝕
corrosion test	腐蚀试验	腐蝕試驗
corrugated bulkhead	槽型舱壁	波形艙壁
corrugated hatchcover	波形舱口盖	波形艙蓋
corvette	轻型护卫舰	輕型巡防艦
counter rudder (= bulb-type rudder)	整流罩舵	整流舵,球形舵
counter weight (= mass-balance weight)	配重	配重
coupling	耦合	偶合
coupling motion	耦合运动	偶合運動
course changing quality	转艏性	航向改變性
course keeping (= course keeping quality)	航向保持性	航向保持
course keeping quality	航向保持性	航向保持
cover	罩壳	蓋
covered electrode	焊条	被覆焊條
covering	包覆	包覆,被覆
COW (= crude oil washing)	原油洗舱	原油洗艙
crack (= crazing)	龟裂	龜裂
cracking test	裂纹试验	破裂試驗
crack sensitivity	裂纹敏感性	裂縫敏感度
craft (= boat)	艇	小艇,船艇
crane ship (= floating crane)	起重船	起重[工作]船
crankcase	曲轴箱	曲柄軸箱
crankcase explosion proof door	曲轴箱防爆门	曲柄軸箱防爆門

英 文 名	祖国大陆名	台湾地区名
crankcase explosion relief door (＝crank-case explosion proof door)	曲轴箱防爆门	曲柄轴箱防爆門
crankcase scavenging	曲轴箱扫气	曲柄轴箱掃氣
crank radius	曲柄半径	曲柄半徑
crankshaft	曲轴	曲[柄]轴
crazing	龟裂	龜裂
creep	蠕变	潛變（材料）
crevice corrosion	缝隙腐蚀	間隙腐蝕，隙間腐蝕
crew room	船员室	船員室
crew space (＝crew room)	船员室	船員室
criteria of maneuverability	操纵性衡准	操縱性準則
criteria of seakeeping qualities	耐波性衡准	耐海性準則
criterion of service numeral	业务衡准数	客船[艙區劃分]基準數
critical angle	临界角	臨界角
critical cavitation number	临界空化数	臨界空化數
critical speed	临界转速	臨界速率，共振轉速
cross bitt	十字带缆桩	十字形系樁
cross curves of stability	形状稳性力臂曲线	[穩度]交叉曲線
cross joint (＝cross shaped joint)	十字接头	十字接頭
cross shaped joint	十字接头	十字接頭
cross structure	连接桥	連橋
crow's nest	了望台	桅桿[了]望台
crucible furnace	坩埚炉	坩堝爐
crude oil carrier (＝crude oil tanker)	原油油船	原油輪
crude oil cleaning system	原油洗舱系统	原油洗艙系統
crude oil tanker	原油油船	原油輪
crude oil washing (COW)	原油洗舱	原油洗艙
crude oil washing system (＝crude oil cleaning system)	原油洗舱系统	原油洗艙系統
cruiser	巡洋舰	巡洋艦
cruiser stern	巡洋舰艉	巡洋艦型艉
cruising ability (＝endurance)	续航力	續航[能]力
cruising speed	巡航速度	巡航船速
CSC (＝International Convention for Safety Container)	国际集装箱安全公约	國際安全貨櫃公約
CSR (＝continuous service rating)	持续常用功率	額定連續常用出力
curing	固化	硬化

英 文 名	祖 国 大 陆 名	台 湾 地 区 名
curing time	硬化时间	硬化時間
current	①海流 ②电流	①洋流 ②電流
curtain plate	檐板	廉板
curved section	曲面分段	
curve of centers of buoyancy	浮心曲线	浮[力中]心曲線
curve of dynamical stability	动稳性曲线	動穩度曲線
curve of extinction	消灭曲线	衰減曲線
curve of static stability (= statical stability curve)	静稳性曲线	靜穩度曲線
cushion	气垫	氣墊
cutter (= small craft)	小艇	小艇
cutter suction dredger (= rock cutter suction dredger)	铰吸式挖石船	鉸刀吸入式挖泥船
cutting line (= line of cut)	切割线	切割線
cutting sequence	切割顺序	切割順序
cutting speed	切削速度	切削速度
cutting torch	割炬	火焰截割器
cycloidal propeller	平旋推进器	擺線推進器,垂直翼螺槳
cyclone separator	旋风分离器	旋風分離器
cylinder	①气缸 ②汽缸	①氣缸 ②汽缸
cylinder block	气缸体	缸體
cylinder bore	气缸直径	缸內徑
cylinder cover (= cylinder head)	气缸盖	缸蓋
cylinder head	气缸盖	缸頭
cylinder jacket (= cylinder liner)	气缸套	缸套
cylinder liner	气缸套	缸襯[套]
cylinder sticking	咬缸	缸膠著
cylinder volume (= working medium volume)	气缸容积	氣缸容積

D

英 文 名	祖 国 大 陆 名	台 湾 地 区 名
daily service fresh water pump	日用淡水泵	日用淡水泵
daily service fuel oil pump	日用燃油泵	日用燃油泵
damage	损伤	損傷
damage control equipment room	损管设备室	

英　文　名	祖国大陆名	台湾地区名
damaged stability	破舱稳性	破損穩度
dan boat	航标艇	設標艇
danger angle（＝critical angle）	临界角	臨界角
dangerous cargo	危险货物	危險貨物
dangerous goods（＝dangerous cargo）	危险货物	危險貨物
datum level（＝base plane）	基面	基準面
datum line（＝base line）	基线	基線,基準線
davit	杂物吊杆	吊柱,吊架
davit launched type liferaft	吊放式救生筏	
davit span	横张索	吊架跨索
daylight signalling light	白昼信号灯	日間訊號燈
D. C. electric propulsion plant	直流电力推进装置	直流電力推進裝置
deadlight	舷窗盖	舷窗內蓋,內窗蓋
dead metal region	死区	死空間
deadrise	舭部升高	［舯］橫斜高
dead soft annealing（＝full annealing）	完全退火	完全退火
dead space（＝dead metal region）	死区	死空間
deadweight	载重量	載重［量］
deadweight scale	载重量标尺	載重標尺
deadwood（＝skeg）	艉鳍	艉鰭
dead zone	盲区	盲區
deaeration	脱气	除氣
deck	甲板	甲板
deck cargo	舱面货	艙面貨,甲板貨
deck center line	甲板中线	甲板中線
deck compression chamber	甲板加压舱	船上加壓室（潛水作業）
deck covering	甲板敷料	甲板被覆
deck equipment and fittings	舱面属具	甲板裝具
deck fittings（＝deck equipment and fittings）	舱面属具	甲板裝具
deck girder	甲板纵桁	甲板縱梁
deck house	甲板室	甲板室
deck ladder	甲板梯	甲板梯
deck light	甲板窗	甲板透光玻璃
deck line	甲板线	甲板線（載重線標誌）
deck longitudinal	甲板纵骨	甲板縱材
deck outfit（＝deck outfitting）	外舾装	艙面艤裝

英 文 名	祖国大陆名	台湾地区名
deck outfitting	外舾装	艙面艤裝
deck paint	甲板涂料	甲板漆
deck runner（＝deck longitudinal）	甲板纵骨	甲板縱材
deck scupper	甲板排水口	甲板排水孔
deck sprinkle system	甲板洒水系统	甲板灑水系統
deck stopper（＝chain stopper）	掣链器	制鏈器（錨）
deck store（＝hawser store）	帆缆间	帆纜庫
deck stringer	甲板边板	甲板緣[厚]板
deck transverse	甲板横桁	甲板深橫桁
deck water seal	甲板水封	甲板水封
deck wetness	甲板淹湿	甲板上浪,甲板濺濕
deep diving submersible	深潜器	深潛器
deep diving system	深潜系统	深海潛水系統
deep draught vessel light	吃水限制灯	吃水限制號燈
deepest subdivision loadline	最深分舱载重线	最深艙區劃分載重線
deep frame（＝deep tank frame）	深舱肋骨	深肋骨
deep penetration welding	深熔焊	深透焊接
deep submergence rescue vehicle（＝deep submersible rescue vehicle）	深潜救生艇	深潛救難艇
deep submersible rescue vehicle（DSRV）	深潜救生艇	深潛救難艇
deep tank	深舱	深艙
deep tank frame	深舱肋骨	深肋骨
deformation	变形	變形
degaussing ship	消磁船	消磁船
degaussing vessel（＝degaussing ship）	消磁船	消磁船
degree of reaction	反动度	反動度（渦輪機）
degree of superheat	过热度	過熱度
degree of vacuum（＝vacuum）	真空度	真空度
delayed crack	延迟裂纹	延遲龜裂
delayed cracking（＝delayed crack）	延迟裂纹	延遲龜裂
delivery ratio	给气比	給氣比
demagnetization	退磁	消磁,去磁
demihull	片体	[雙體船的]半[船]體
deoxidizer	脱氧剂	除氧劑,還原劑
depolarization	去极化	去極化作用
deposited metal	熔敷金属	堆積金屬,溶著金屬（焊接）
deposition efficiency	熔敷效率	澱積效率,堆積效率,溶

英 文 名	祖 国 大 陆 名	台 湾 地 区 名
		著效率（焊接）
deposition metal（＝deposited metal）	熔敷金属	堆積金屬,溶著金屬（焊接）
deposition rate	①沉积速度 ②熔敷速度	①澱積率,堆積率 ② 溶著速度
depth charge	深水炸弹	深水炸彈
depth of fusion（＝penetration）	熔深	熔解深度
depth sounder	测深	測深
derrick（＝derrick rig）	吊杆装置	吊桿裝置
derrick barge（＝floating crane）	起重船	起重［工作］船
derrick boom	吊杆	吊桿,起重桿
derrick mast	起重桅	吊桿桅
derrick post	起重柱	吊桿柱
derrick rest	吊杆托架	吊桿托架
derrick rig	吊杆装置	吊桿裝置
derrick rigging	起货索具	吊桿索具
derrick socket（＝gooseneck bracket）	吊杆座	吊桿承座
design datum	设计基准	設計基準（準則）
designed displacement	设计排水量	設計排水量
designed draft	设计吃水	設計吃水
designed waterline	设计水线	設計水線
design load	计算载荷	設計負荷
design norm（＝design datum）	设计基准	設計基準（準則）
design speed	设计航速	設計船速
destroyer	驱逐舰	驅逐艦
destruction test（＝destructive test）	破坏性试验	破壞性試驗
destructive test	破坏性试验	破壞性試驗
detector for fire alarm system	火灾报警探测器	火警探測器
determination for main engine location	主机校中	
developed area	展开面积	展開面積
devil's claw	掣链钩	吊鏈鈎,拉線爪
dew point	露点	露點
diagonal	斜剖线	斜剖面線
diagonal engine（＝inclined engine）	斜置式内燃机	斜置引擎
diagonal line（＝diagonal）	斜剖线	斜剖面線
diameter factor（＝Taylor's diameter constant）	直径系数	直徑因數（螺槳）
diaphragm	隔膜,保护膜	隔膜,膜片

英　文　名	祖 国 大 陆 名	台 湾 地 区 名
die casting	压力铸造	壓鑄,模鑄
die forging	模锻	模鍛
dielectric strength test（=dielectric test）	介质强度试验	介質強度試驗
dielectric test	介质强度试验	介質強度試驗
diesel boat（=diesel ship）	柴油机船	柴油機船,內燃機船
diesel dynamo（=diesel generator）	柴油发电机	柴油發電機
diesel-electric propulsion plant	柴油机电力推进装置	
diesel engine	柴油机	柴油引擎,狄塞爾引擎
diesel engine power plant	柴油机动力装置	
diesel generating set	柴油发电机组	柴油發電機組
diesel generator	柴油发电机	柴油發電機
diesel oil daily tank	柴油日用柜	柴油日用櫃
diesel oil settling tank	柴油沉淀柜	柴油沉澱櫃
diesel ship	柴油机船	柴油機船,內燃機船
diffraction force	绕射力	繞射[波]力
diffuser	①扩压器 ②扩散口	擴散器
diffuser pump	导叶泵	擴散泵
dimension ratio	主尺度比	
dip brazing	浸沾钎焊	熱浸硬焊
dipper	铲斗	杓
dipper crowding gear	推压机构	推壓機構
dipper dredger	正铲挖泥船	戽斗挖泥船
dipper machine	铲扬机	鏟揚機
dipping	艉沉	艉俯
direct current generator	直流发电机	直流發電機
direct current motor	直流电动机	直流馬達
directional spectrum	方向谱	方向頻譜
directional stability	方向稳定性	方向穩定性
direct numerical control（DNC）	直接数控	直接數值控制
direct tower（=command tower）	指挥塔台	指揮塔
dirty ballast	污压载水	不潔壓艙水
dirty tanker（=crude oil tanker）	原油油船	原油輪
disc	轮盘	圓盤
disc clutch	片式离合器	盤形離合器
disc oil separator	碟式油分离机	
discontinuity	不连续[性]	不連續性
displacement	排水量	排水量
displacement margin	储备排水量	排水量餘裕

英 文 名	祖 国 大 陆 名	台 湾 地 区 名
displacement ship	排水型船	排水型船
distilled water tank	蒸馏水柜	蒸餾水櫃
distiller	蒸馏器	蒸餾器
distilling tank (= distilled water tank)	蒸馏水柜	蒸餾水櫃
distribution board	分配电箱	配電盤
distribution box (= distribution board)	分配电箱	分電箱
diverging shafting	外斜轴系	[相對於中心面]外斜轴系（多軸系船）
divided flow turbine (= double-flow steam turbine)	双流汽轮机	雙流渦輪機
diving apparatus	潜水装具	潛水器具
diving bell	潜水钟	潛水鍾
diving boat (= diving support vessel)	潜水工作船,潜水支援船	潛水工作船
diving suit	潜水服	潛水服,潛水衣
diving support vessel	潜水工作船,潜水支援船	潛水支援船
division plate	隔板	隔板
DNC (= direct numerical control)	直接数控	直接數值控制
dock	船坞	塢
dock bottom	坞底	塢底
dock chamber	坞室	塢室
dock entrance	坞口	
dock floor (= dock bottom)	坞底	塢底
dock gate (= caisson)	坞门	塢門
dock gate channel	门槽	門槽
dock head	坞首	塢首
docking	进坞	進塢
docking block	坞墩	塢墩
docking keel	坞龙骨	駐塢龍骨
docking repair	坞修	塢修
docking strength	坐坞强度	坐塢強度
dock outfitting	船坞舾装	
dock repair (= docking repair)	坞修	塢修
dock sill	坞坎	塢檻
donkey boiler (= auxiliary boiler)	辅锅炉	輔鍋爐,副鍋爐
Doppler log	多普勒计程仪	都卜勒測程儀
Doppler sonar	多普勒声呐	都葡勒聲納

英　文　名	祖 国 大 陆 名	台 湾 地 区 名
double acting pump	双作用泵	雙動泵
double bearing rudder	双支承舵	雙支承舵
double bottom	双层底	［二］重底
double bottom tank	双层底舱	［二］重底艙
double bulkhead (= double plate bulk- head)	双板舱壁	雙層艙壁
double-flow steam turbine	双流汽轮机	雙流渦輪機
double flow turbine (= double-flow steam turbine)	双流汽轮机	雙流渦輪機
double groove	双面坡口	雙面槽
double hulled ship (= double-skin ship)	双壳船	雙殼船
double-leaf door	双截门	雙截門
double model	重叠船模	對疊［船］模,重疊模
double plate bulkhead	双板舱壁	雙層艙壁
double-skin ship	双壳船	雙殼船
doubling plate	覆板	加力板,加強板,二重板
downcomer	下降管	降水管（鍋爐）
downcomer tube (= downcomer)	下降管	降水管（鍋爐）
draft	吃水	吃水
draft loss	通风阻力	通風損失
draft mark	吃水标志	吃水標
drag	拖曳	拖曳
drag chain	牵引链条	制動鏈（下水）
drag link	拉杆	拉桿
drain	疏水	排泄
drainage pump	疏水泵	污水泵
drain cock	泄放旋塞阀	排泄旋塞
drain hole	流水孔	排泄孔
drain pipe	疏水管	排泄管
drain pump (= drainage pump)	疏水泵	污水泵
drain tank	疏水箱	排泄櫃
drain valve	疏水阀	排泄閥
draught (= draft)	吃水	吃水
draught forward (= fore draft)	艏吃水	艏吃水
draught mark (= draft mark)	吃水标志	吃水標
dredge pump	泥泵	挖泥泵
dredger	挖泥船	挖泥船
drift	横漂	漂流,飄移

英 文 名	祖国大陆名	台湾地区名
drifter（＝drift netter）	流[刺]网渔船	流網渔船
drift fishing boat（＝drift netter）	流[刺]网渔船	流網渔船
drift netter	流[刺]网渔船	流網渔船
drilling	钻削	鑽鑿
drilling barge	钻井驳	鑽油駁
drilling platform	钻井平台	鑽油平台
drilling rig（＝drilling platform）	钻井平台	鑽油台
drilling ship	①钻井船 ②钻探船	鑽油船
drilling tender	钻井供应船	鑽油台補給船
drilling unit（＝drilling platform）	钻井平台	鑽油台
drilling vessel（＝drilling ship）	①钻井船 ②钻探船	鑽油船
drinking water tank	饮水舱	飲用水艙
drive on/drive off ship（＝roll on-roll off ship）	滚装船	滚装船,駛上駛下船,轆轆船
drop hammer	单作用锤	落錘
droppable ballast	可弃压载	可棄壓載
dropping	艏落	艏落
drop test	坠落试验	墜落試驗
drum	卷筒	捲索鼓,鼓輪
dry-card compass	干罗经	乾羅經
dry cargo ship	干货船	乾貨船
dry compass（＝dry-card compass）	干罗经	乾羅經
dry dock	干船坞	乾[船]塢
DSRV（＝deep submersible rescue vehicle）	深潜救生艇	深潛救難艇
dual-fuel engine	双燃料发动机	雙燃料引擎
duct center girder	箱形中桁材	箱形中線縱梁
ducted propeller	导管推进器	導罩螺槳
dummy model	假模	假模
dummy piston	平衡活塞	均壓活塞
dummy propeller boss	假毂	假轂
duplex pump	双联泵	雙缸泵
duplicate supply（＝two circuit feeding）	双路供电	雙路供電,雙重供電
duration	工作持续时间	歷時,時段
dust-protected	防尘	防塵
dynamical heeling angle	动横倾角	動橫傾角
dynamical stability	动稳性	動穩度
dynamical upsetting angle	动倾覆角	動傾覆角

英　文　名	祖国大陆名	台湾地区名
dynamic balancing	动平衡	動態平衡
dynamic balancing test	动平衡试验	動平衡試驗
dynamic load	动载荷	動態負荷
dynamic positioning	动力定位	動態定位
dynamic stability	动稳定性	動穩度
dynamo engine（=genset engine）	发电用内燃机	發電機引擎
dynamotor	直流电动发电机	電動發電機

E

英　文　名	祖国大陆名	台湾地区名
earth（=earthing）	接地	接地
earth bar（=earth electrode）	接地极	接地棒
earth connection（=earthing）	接地	接地
earthed circuit	接地电路	接地回路
earth electrode	接地极	接地棒
earthing	接地	接地
eccentricity	偏心度	偏心度,偏心距
echo	回波	回波（電）
echo sounder（=echo sounding）	回声测深仪	回聲測深儀
echo sounding	回声测深仪	回聲測深儀
echo sounding machine（=echo sounding）	回声测深仪	回聲測深儀
economical power	经济功率	經濟功率
economical speed	经济转速	經濟船速
eddy making damping	旋涡阻尼	旋渦阻尼
edge	端面	邊緣
edge preparation	边缘加工	邊緣[預]加工
edge weld	端接焊缝	邊[緣]焊接
effective freeboard	有效干舷	有效乾舷
effective power	有效功率	有效功率
effective range	有效射程	有效射程
effective thermal efficiency（=effective brane thermal efficiency）	有效热效率	有效熱效率
effective thrust	有效推力	有效推力
effective wake	实效伴流	有效跡流
effective wave slope	有效波倾角	有效波斜度
elastic coupling（=resilient shaft coup-	弹性联轴器	彈性聯軸器

英　文　名	祖 国 大 陆 名	台 湾 地 区 名
ling		
electrical power	电功率	電功率
electrical shop（＝electrician's store）	电工间	電工場
electric engine telegraph	电动主机传令装置	電動車鐘
electric fitting	电装	電裝
electric heater	电加热器	電熱器
electric heating	电加热	電［加］熱
electrician's store	电工间	電工場
electric motive force（＝electromotive for-ce）	电动势	電動勢
electric power（＝electrical power）	电功率	電功率
electric propulsion	电力推进	電力推進
electric propulsion plant	电力推进装置	電力推進設備
electric propulsion ship	电力推进船	電力推進船
electric rudder angle indicator	舵角指示器	舵角指示器
electric steering gear	电动操舵装置	電動舵機
electric test	电气试验	電試驗
electrochemical corrosion	电化学腐蚀	電化學腐蝕
electrode	电极	電極
electro-hydraulic steering gear	电动液压操舵装置	電動液壓操舵系統
electro-hydraulic steering system（＝elec-tro-hydraulic steering gear）	电动液压操舵装置	電動液壓操舵系統
electrolyte	电解质	電解質
electromagnet	电磁铁	電磁鐵
electromagnet control gyrocompass	电控罗经	
electromagnetic clutch	电磁离合器	電磁離合器
electromagnetic compatibility（EMC）	电磁兼容性	電磁相容性
electromagnetic interference（EMI）	电磁干扰	電磁干擾
electromagnetic log	电磁计程仪	電磁計程儀
electromotive force	电动势	電動勢
electron beam welding	电子束焊	電子束焊
electronic governor	电子调速器	電子調速器
electroplating	电镀	電鍍
electro-slag welding	电渣焊	電熱熔渣焊接
elliptical stern	椭圆艉	橢圓艉
embarkation ladder	救生登乘梯	［艇筏］承載梯
embarkation lamp（＝boat deck light）	登艇灯	［艇筏］乘載照明燈,小艇甲板燈

英　文　名	祖国大陆名	台湾地区名
EMC（＝electromagnetic compatibility）	电磁兼容性	電磁相容性
emergency automatic shut down device（＝emergency shutdown system）	应急关闭系统	應急自動關閉裝置
emergency blower（＝emergency standby blower）	应急辅助鼓风机	應急鼓風機
emergency draining test	应急疏水试验	應急泄水試驗
emergency electrical power plant	应急电站	應急電站
emergency electrical power source（＝emergency source of electrical power）	应急电源	應急電源
emergency fire pump	应急消防泵	應急消防泵
emergency generator	应急发电机	應急發電機
emergency light	应急灯	應急燈
emergency shutdown system	应急关闭系统	應急自動關閉裝置
emergency source of electrical power	应急电源	應急電源
emergency standby blower	应急辅助鼓风机	應急鼓風機
emergency steering gear	应急操舵装置	應急操舵裝置,應急舵機
EMI（＝electromagnetic interference）	电磁干扰	電磁干擾
emulsification	乳化	乳化
emulsified fuel	乳化燃料	乳化燃料
enclosed spaces	围蔽处所	圍蔽空間
enclosure	外壳	外殼
end launching	纵向下水	艉向下水
end shackle	末端卸扣	端接環
end slipway	纵向滑道	縱向滑道
endurance	续航力	續航[能]力
endurance test	耐久[性]试验	持久試驗
engine aft（＝stern-engined ship）	艉机型船	艉機艙船,艉機型船
engine bed（＝bed plate）	机座	機座
engine block	机体	機體,引擎體
engine compartment coating（＝engine compartment paint）	机舱涂料	機艙塗料
engine compartment paint	机舱涂料	機艙塗料
engine control room	机舱集控室	機艙控制室
engine cut off test	减机运转试验	停機試驗
engine room	机舱	機艙
engine room arrangement	机舱布置	機艙佈置
engine room automation	机舱自动化	機艙自動化

英　文　名	祖国大陆名	台湾地区名
engine-room auxiliary machinery	机舱辅机	
engine room casing	机舱棚	機艙棚,機艙天罩
engine seat（=bed plate）	机座	機座
enlarged link	加大链环	加大鏈環
entrance	进流段	艏入水段
entry locker（=transfer chamber）	过渡舱	過渡艙
environmental load	环境载荷	環境負荷
environmental monitoring ship（=environmental monitoring vessel）	环境监测船	
environmental monitoring vessel	环境监测船	
equipment for collision avoidance	避碰装置	
equipment number	舾装数	舾裝數,屬具數
erosion	溶蚀	沖蝕,潰蝕
escape hatch	应急舱口	逃生艙口
escape scuttle	脱险口	逃生窗口
escort boat	护卫艇	護航艇
examination of opened up parts（=open-up examination）	拆验	拆檢
excess air factor（=excess air ratio）	［燃烧］过量空气系数	過量空氣因數
excess air ratio	［燃烧］过量空气系数	過量空氣因數
exchanging water hole	换水孔	換水孔
exciter	励磁机	勵磁機,激磁機
exciter test	激振试验	激振試驗
exciting force	扰动力	激振力
excluded spaces	免除处所	免［除］丈［量］空間
excursion boat（=excursion vessel）	旅游船,游览船	遊覽船
excursion vessel	旅游船,游览船	遊覽船
Exemption Certificate	免除证书	豁免證書
exhaust duct	排气道	排氣導管
exhaust fan	排气风机	排氣機,抽風機
exhaust steam	排汽	排汽
exhaust steam system	排汽系统	排汽系統
exhaust steam turbine	乏汽轮机	排［蒸］汽渦輪機
exhaust stroke	排气行程	排氣衝程
exhaust system（=exhaust steam system）	排汽系统	排汽系統
exhaust unit	排气装置	排氣裝置
expanded area	伸张面积	伸展面積（螺槳）
expansion joint	①伸缩接管 ②伸缩接	①伸縮接合 ②伸縮接

英　文　名	祖国大陆名	台湾地区名
	头	頭
expansion stroke	膨胀行程	膨脹衝程
expansion [pressure] ratio	膨胀[压]比	膨脹比
explosion welding	爆炸焊	爆炸焊接
exposed core (=bare terminal)	焊条夹持端	焊條裸端
extinguisher (=fire extinguishers)	灭火器	滅火器
extraction turbine	抽汽式汽轮机	抽汽渦輪機
extreme breadth	最大宽	最大寬度,全寬
extreme length	最大长	全長
eye brow	眉毛板	窗楣
eye plate	眼板	眼板

F

英　文　名	祖国大陆名	台湾地区名
fabrication	分段装配	[船體]分段裝配
FAC(missile) (=guided missile boat)	导弹快艇	[導向]飛彈快艇
face cavitation	叶面空化	[螺槳]葉[正]面空化
face plate	面板	面板
factor of subdivision	分舱因数	艙區劃分因數
factory trawler	加工拖网渔船	拖網加工船
failure	故障	故障
failure analysis	故障分析	故障分析
failure mode	失效模式	故障模式
failure mode and effect analysis (FMEA) (=failure mode effect and criticality analysis)	失效模式、效应及后果分析	故障模式及影響分析
failure mode effect and criticality analysis	失效模式、效应及后果分析	故障模式及影響分析
failure rate	失效率	失效率,故障率
fairlead	导缆器	導索器
fairleader (=fairlead)	导缆器	導索器
false bottom	假底	[水槽]活動底
fan room	通风机室	風扇室
fatigue	疲劳	疲勞
fault current	故障电流	故障電流
faying surface	贴合面	接合面
feathering paddle wheel (=paddle wheel)	明轮	明輪

英 文 名	祖 国 大 陆 名	台 湾 地 区 名
feathering step（＝feathering tread）	活动梯步	舷梯活動踏步
feathering tread	活动梯步	舷梯活動踏步
feed	进给量	供給
feeder panel	配电屏	饋電盤
feed valve（＝feed water valve）	给水阀	給水閥
feed water	给水	給水,爐水
feed water heater	给水加热器	給水加熱器
feed water valve	给水阀	給水閥
fender	①护舷材 ②碰垫	①護舷材 ②碰墊
ferrocement boat	钢丝网水泥船	鋼筋水泥船
ferrocement ship（＝ferrocement boat）	钢丝网水泥船	鋼筋水泥船
ferrocement vessel（＝ferrocement boat）	钢丝网水泥船	鋼筋水泥船
ferry	渡船	渡船
ferry boat（＝ferry）	渡船	渡船
FFG（＝guided missile frigate）	导弹护卫舰	飛彈巡防艦
fiberglass reinforced plastic boat（＝fiberglass reinforced plastic ship）	玻璃钢船	玻[璃]纖[維]強化塑膠船
fiberglass reinforced plastic ship（FRP ship）	玻璃钢船	玻[璃]纖[維]強化塑膠船
fiberglass reinforced plastics（FRP）（＝glass fiber reinforced plastics）	玻璃纤维增强塑料	玻[璃]纖[維]強化塑膠
fiber rope	纤维索	纖維纜線
field rheostat	励磁变阻器	磁場變阻器
field welding	工地焊接	現場焊接
filler	填充物	填料,填充劑
fillet weld	角焊缝	填角焊道
fillet weld in normal shear	正面角焊缝	正面填角焊道
fillet weld in parallel shear	侧面角焊缝	側面填角焊接
fillet welding	角焊	填角焊[接]
filling pipe（＝filling piping）	注入管	注入管
filling piping	注入管	注入管
film	涂膜	塗膜
filter	滤器	過濾器
filter glass	黑玻璃	濾光玻璃
fin	鳍	鰭板
final blade	末叶片	最後葉片
final inspection	完工检验	完工檢查
fire alarm（＝fire alarm sounder）	火灾警报器	火警警報器

英　文　名	祖国大陆名	台湾地区名
fire alarm sounder	火灾警报器	火警警報器
fire boat (＝firefighting ship)	消防船	消防艇
fire box	火箱	火箱
firebrick	耐火砖	耐火磚,火磚,燒磨土磚
fireclay	耐火黏土	耐火泥,火泥,燒磨土
fire combustion chamber (＝combustion chamber)	燃烧室	燃燒室
fire-control room	消防控制室	火警控制室
fire damper	防火风门	防火擋板
fire detector (＝detector for fire alarm system)	火灾报警探测器	火警探測器
fire door	防火门	防火門
fire extinguishers	灭火器	滅火器
firefighting ship	消防船	消防艇
fire point	燃点	燃點
fireproof bulkhead	防火舱壁	防火艙壁
fire protected lifeboat (＝fire-resistant lifeboat)	耐火救生艇	防火救生艇
fire pump	消防泵	消防泵,滅火泵
fire-resistant lifeboat	耐火救生艇	防火救生艇
fire resisting cables	耐火电缆	抗燃電纜
fire room (＝boiler room)	锅炉舱	鍋爐艙,鍋爐間
fire-tube boiler	烟管锅炉	火管鍋爐
firewall (＝fireproof bulkhead)	防火舱壁	防火艙壁
firing order	发火次序	點火順序
fisheries factory ship (＝fish factory ship)	渔业加工船	漁業加工船
fisheries guidance boat (＝fishery guidance ship)	渔业指导船	漁業指導船
fisheries research boat (＝fishery research vessel)	渔业调查船	漁業研究船
fisheries training boat (＝fishery training vessel)	渔业实习船	漁業訓練船
fishery factory ship (＝fish factory ship)	渔业加工船	漁業加工船
fishery guidance ship	渔业指导船	漁業指導船
fishery research vessel	渔业调查船	漁業研究船
fishery training vessel	渔业实习船	漁業訓練船
fish factory ship	渔业加工船	漁[獲]加工船
fish finder	鱼探仪	魚[群]探[測]儀

英 文 名	祖 国 大 陆 名	台 湾 地 区 名
fish hold	鱼舱	魚艙
fishing boat（=fishing vessel）	渔船	漁船
fishing light	渔业灯	捕魚燈
fishing machinery	捕捞机械	漁撈機械
fishing port supervision boat	渔港监督艇	
fishing vessel	渔船	漁船
fish luring light boat	灯光诱鱼船	誘魚燈船
fitting-out quay	舾装码头	
fixture	焊接夹具	夾具
flag chest	旗箱	旗箱
flag locker（=flag chest）	旗箱	旗箱
flame arrester	防火网	減焰器
flame cleaning	火焰除锈	火焰清除
flame detector	火焰监测器	火焰探測器
flame gouging	火焰气刨	火焰開槽
flame hardening	火焰淬火	火焰硬化
flame tube	火焰管	火管
flanged bracket	折边肘板	卷邊腋板,凸緣腋板
flanged coupling	凸缘联轴器	凸緣連接器
flange joint	法兰连接	凸緣接合
flap	襟翼	襟翼
flap rudder（=flap-type rudder）	襟翼舵	襟翼舵
flap-type rudder	襟翼舵	襟翼舵
flare	外倾	舷緣外傾
flash（=flashing）	闪光	閃光
flashback（=back fire）	①回火 ②逆火	①回火 ②逆火
flash butt welding	闪光对焊	閃電對頭焊接
flashing	闪光	閃光
flash point	闪点	閃點,引火點
flat position welding	平焊	平焊,平焊接
flat section	平面分段	
flat welding（=flat position welding）	平焊	平焊,平焊接
flexibility	柔韧性	柔性,柔度
flexible cable	软电缆	柔性電纜,柔性纜
flexible manufacturing cells（FMC）	柔性制造单元	
flexible manufacturing system（FMS）	柔性制造系统	彈性製造系統
flexible rotor	挠性转子	柔性轉子
flight deck	飞行甲板	飛行甲板（航艦）

英　文　名	祖国大陆名	台湾地区名
floated gyro	液浮陀螺仪	液體羅經
floating anchor	浮锚	海錨
floating body	浮体	浮體
floating concrete mixer	混凝土搅拌船	
floating condition	浮态	浮揚狀態
floating crane	起重船	起重[工作]船
floating dock	浮船坞,浮坞	浮塢
floating launching	漂浮式下水	漂浮下水
floating on even keel	正浮	縱平浮
floating pile driver	打桩船	打樁船
floating pile-driving plant (=floating pile driver)	打桩船	打樁船
floating production storage offloading (FPSO)	浮式生产储卸油装置	浮式採油儲油及卸油
floating production storage unit (FPSU)	浮式生产储油装置	浮式生産儲油單元
floating production system (FPS)	浮式生产系统	浮式生産系統
float level gauge	浮筒液位计	浮動液位計
floodable length	可浸长度	可浸長度
flooded waterline (=flood waterline)	破舱水线	浸水吃水線
flooding angle	进水角	浸水角,泛水角
flooding dock	注水式船坞	
flooding of pressure hull	耐压壳体进水	
flood waterline	破舱水线	浸水吃水線
floor	肋板	底肋板
flow coefficient	流量系数	流量係數
flow meter	流量计	流量計
flow pattern	流型	流型
flow visualization	流动显示	流場觀察,流場可視化
flue (=gas pass)	烟道	煙道
flue gas analysis	烟气分析	煙道氣分析
fluid compass (=floated gyro)	液浮陀螺仪	液體羅經
fluidized bed boiler	流化床锅炉	流動床[燃燒]鍋爐
fluidized bed firing boiler (=fluidized bed boiler)	流化床锅炉	流動床[燃燒]鍋爐
fluorescence	荧光	熒光[性]
fluorescent screen	荧光屏	熒光屏
fluoroscope (=fluorescent screen)	荧光屏	熒光屏
flush deck ship	平甲板船	平甲板船

英 文 名	祖 国 大 陆 名	台 湾 地 区 名
flush deck vessel（＝flush deck ship）	平甲板船	平甲板船
flushing of pipeline	管路冲洗	
flush weld（＝flat position welding）	平焊	平焊,平焊接
flux combined wire（＝flux-cored wire）	药芯焊丝	含焊劑芯焊線,複合焊線
flux-cored wire	药芯焊丝	含焊劑芯焊線,複合焊線
flying-off deck（＝flight deck）	飞行甲板	飛行甲板（航艦）
fly wheel	飞轮	飛輪
FMC（＝flexible manufacturing cells）	柔性制造单元	
FMEA（＝failure mode and effect analysis）	失效模式、效应及后果分析	故障模式及影響分析
FMS（＝flexible manufacturing system）	柔性制造系统	彈性製造系統
foam fire extinguisher	泡沫灭火器	泡沫滅火器
foam fire extinguishing system	泡沫灭火系统	泡沫滅火系統
foamite extinguisher（＝foam fire extinguisher）	泡沫灭火器	泡沫滅火器
folding hatchcover	折叠式舱口盖	折疊[式]艙口蓋
following sea	随浪	順浪,從浪
foot steps	踏步	踏板
forced circulation boiler	强制循环锅炉	強制循環鍋爐
fore body	前体	艏部
forecastle	艏楼	艏樓
fore draft	艏吃水	艏吃水
forefoot	艏踵	艏材跟部
forehand welding	左焊法	前進焊法,[右手]左向焊法
fore mast	前桅	前桅
fore peak	艏尖舱	艏尖艙,前尖艙
forepeak pump	艏尖舱泵	艏尖艙手動泵
forepeak tank（＝fore peak）	艏尖舱	艏尖艙,前尖艙
fore perpendicular（＝forward perpendicular）	艏垂线	艏垂標
fore poppet	艏支架	艏支架,艏托台（下水）
forge weld（＝forging welding）	锻接	鍛接
forging hammer	锻锤	鍛錘
forging ratio	锻造比	鍛造比
forging welding	锻接	鍛接

英　文　名	祖国大陆名	台湾地区名
fork beam (= hatch side cantilever beam)	舱口悬臂梁	艙口側半梁
form coefficient	形状系数	形狀係數
forming	成形法	成型法
forward draft (= fore draft)	艏吃水	艏吃水
forward draught (= fore draft)	艏吃水	艏吃水
forward perpendicular	艏垂线	艏垂標
foundry	铸造	鑄造
four cycle (= four-stroke cycle)	四冲程循环	四衝程循環
four cycle engine (= four-stroke engine)	四冲程内燃机	四衝程引擎
four-stroke cycle	四冲程循环	四衝程循環
four-stroke engine	四冲程内燃机	四衝程引擎
FPS (= floating production system)	浮式生产系统	浮式生產系統
FPSO (= floating production storage off-loading)	浮式生产储卸油装置	浮式採油貯油及卸油
FPSU (= floating production storage unit)	浮式生产储油装置	浮式生產儲油單元
frame	肋骨	肋骨
frame ring	肋骨框架	肋骨圈
frame space (= frame spacing)	肋距	肋骨間距
frame spacing	肋距	肋骨間距
framing	骨架	骨架,構架
freeboard	干舷	乾舷
freeboard deck	干舷甲板	乾舷甲板
free fall lifeboat	抛落式救生艇	
free piston air compressor	自由活塞空气压缩机	自由活塞壓縮機
free piston compressor (= free piston air compressor)	自由活塞空气压缩机	自由活塞壓縮機
free-running model test	自由自航船模试验	自航船模[運動]試驗
free surface	自由液面	自由液面
free surface correction	自由液面修正	自由液面修正
freighter (= cargo ship)	货船	貨船
frequency changer (= frequency convertor)	变频器	變頻器,變頻機
frequency convertor	变频器	變頻器,變頻機
fresh water cooler	淡水冷却器	淡水冷卻器
fresh water filter	淡水滤器	淡水過濾器
fresh water heater	淡水加热器	淡水加熱器
fresh water pump	淡水泵	淡水泵
fresh water tank	淡水舱	淡水艙,淡水櫃

英　文　名	祖国大陆名	台湾地区名
fretting corrosion	摩振腐蚀	磨[耗腐]蚀
frictional damping	摩擦阻尼	摩擦阻尼
frictional resistance	摩擦阻力	摩擦阻力
friction clutch	摩擦式离合器	摩阻離合器
friction welding	摩擦焊	摩擦焊
frigate	护卫舰	巡防艦
front bulkhead	前端壁	前端壁
front fillet weld（=fillet weld in normal shear）	正面角焊缝	正面填角焊道
FRP（=fiberglass reinforced plastics）	玻璃纤维增强塑料	玻[璃]纖[維]強化塑膠
FRP ship（=fiberglass reinforced plastic ship）	玻璃钢船,玻璃纤维增强塑料船	玻[璃]纖[維]強化塑膠船
fuel coefficient	燃料系数	耗油係數
fuel consumption	①燃油消耗量 ②燃料消耗量	① 燃油消耗量 ②燃料消耗量
fuel consumption test	燃料消耗试验	燃料消耗試驗
fuel oil burning pump	锅炉燃油泵	鍋爐燃油泵
fuel oil daily tank	燃油日用柜	燃油日用櫃
fuel oil drain system	燃油泄放系统	燃油泄放系統
fuel oil filter	燃油滤器	燃油過濾器
fuel oil settling tank	燃油沉淀柜	燃油沉澱櫃
fuel oil system	燃油系统	燃油系統
fuel oil tank	①燃油柜 ②燃油舱	①燃油櫃 ②燃油艙
fuel oil transfer pump	燃料油输送泵	燃油輸送泵
full annealing	完全退火	完全退火
full load	全负荷	全負荷
full load displacement	满载排水量	滿載排水量
full scale lofting	实尺放样	實尺放樣
fumigation	熏舱	熏艙消毒
function	功能	功能
fundamental frequency	基频	基[本]頻[率],最低頻率,一階頻率
furnace	炉膛	爐膛
furnace brazing	炉中钎焊	爐內硬焊
furnace volume	炉膛容积	爐內容積
fuse	熔断器	熔斷器
fusion welding	熔焊	熔接

G

英 文 名	祖国大陆名	台湾地区名
gain	增益	增益
gallery	廊道	廊,通道
galvanic anode protection	牺牲阳极阴极保护	犧牲陽極陰極保護
galvanic corrosion	电偶腐蚀	電蝕
galvanization (= galvanizing)	热镀锌	鍍鋅
galvanizing	热镀锌	鍍鋅
gangway ladder (= accommodation ladder)	舷梯	舷梯
gangway light	舷门灯	梯口燈
gangway port	舷墙门	梯口通道
garbage boat	垃圾船	垃圾艇
garbage chute	垃圾运送槽	垃圾斜槽
garbage cleaning vessel	垃圾清扫船	
garbage shoot (= garbage chute)	垃圾运送槽	垃圾斜槽
garboard strake	龙骨翼板	龍骨翼板列,A 板列
gas concentration measurement instrument	气体浓度测量仪	氣體濃度測量儀
gas cutting	气割	焰割
gas freeing	驱气系统	貨油艙清除油氣裝置
gas-freeing	驱气	清除[有害]氣體
gas generator	燃气发生器	氣體發生器
gas metal arc welding (GMAW) (= gas shielded arc welding)	气体保护[电弧]焊	氣體遮護金屬電[弧]焊[接]
gasoline engine	汽油机	汽油機,汽油引擎
gas pass	烟道	煙道
gas shielded arc welding	气体保护[电弧]焊	氣體遮護金屬電[弧]焊[接]
gas turbine	燃气轮机	燃氣渦輪機
gas turbine power plant	燃气轮机动力装置	
gas turbine ship	燃气轮机船	燃氣渦輪機船
gas welding	气焊	氣焊
gate chamber	门库	閘門室
gate riser (= riser)	冒口	冒口（鑄造）
general alarm	全船报警装置	

英 文 名	祖 国 大 陆 名	台 湾 地 区 名
general arrangement	总布置	總佈置［圖］
general cargo ship	杂货船	雜貨船
general corrosion	普遍腐蚀	全面腐蝕
general purpose ship（＝multipurpose cargo ship）	多用途货船	多用途貨船
general service pump	总用泵	通用泵,常用泵
generating line	母线	母線
generating set	发电机组	發電機組
generator	发电机	發電機
generator-motor system	发电机电动机系统	列氏電動操作系統
genset engine	发电用内燃机	發電機引擎
geological survey ship	地质调查船	地質調查船
geological survey vessel（＝geological survey ship）	地质调查船	地質調查船
geometrically similar ship models	几何相似船模	幾何相似模型
geometrically similar model（＝geometrically similar ship models）	几何相似船模	幾何相似模型
geophysical survey ship（＝geophysical survey vessel）	地球物理勘探船	地球物理探測船
geophysical survey vessel	地球物理勘探船	地球物理探測船
gill netter	刺网渔船	刺網漁船
gimbal mounting（＝gimbal table）	万向架	萬向吊架,恒平框架
gimbal table	万向架	萬向吊架,恒平框架
girder	桁材	桁,縱梁
glass fiber reinforced plastics（GRP）	玻璃纤维增加塑料	玻［璃］纖［維］強化塑膠
glider（＝planing boat）	滑行艇	滑行［快］艇,水上快艇,滑航艇
global maritime distress and safety system（GMDSS）	全球海上遇险和安全系统	全球海上遇險及安全系統
global positioning system（GPS）	全球定位系统	全球［衛星］定位系統
global vibration	总振动	
globular transfer	粗滴过渡	粒狀傳遞
globule	熔滴	熔滴（電焊）
GMAW（＝gas metal arc welding）	气体保护［电弧］焊	氣體遮護金屬電［弧］焊［接］
GMDSS（＝global maritime distress and safety system）	全球海上遇险和安全系统	全球海上遇險及安全系統

英　文　名	祖国大陆名	台湾地区名
gooseneck bracket	吊杆座	鹅颈型吊桿座
governor	调速器	调速器
GPS（＝global positioning system）	全球定位系统	全球[衛星]定位系统
grab	抓斗	抓斗
grab dredger	抓斗挖泥船	抓斗[式]挖泥船
grab line（＝grab stabilizer line）	抓斗稳索	抓索
grab stabilizer line	抓斗稳索	抓索
grain size	晶粒度	粒度
grapple dredger（＝grab dredger）	抓斗挖泥船	抓斗[式]挖泥船
graving dock（＝dry dock）	干船坞	乾[船]坞
gravity launching	重力式下水	重力下水
gravity oil tank	重力油柜	重力油櫃
gravity platform	重力式平台	重力式鑽油台
gravity type boat davit（＝gravity-type davit）	重力式吊艇架	重力[小艇]吊架
gravity-type davit	重力式吊艇架	重力[小艇]吊架
gravity welding	重力焊	重力式[電弧]焊接
gray cast iron	灰铸铁	灰[口]生鐵
gray pig iron（＝gray cast iron）	灰铸铁	灰[口]生鐵
grid	格子线	格子
gripgear（＝jigs and fixtures）	夹具	夾具,緊固具
groove angle	坡口角度	開槽角
groove face	坡口面	開槽面
groove radius（＝root radius）	根部半径	根[部]半徑
grooving corrosion（＝groovy corrosion）	沟状腐蚀	槽蝕
groovy corrosion	沟状腐蚀	槽蝕
gross tonnage	总吨位	總噸位
grounding（＝earthing）	接地	接地
ground resistance（＝resistance of an earthed conductor）	接地电阻	接地電阻
ground return circuit（＝earthed circuit）	接地电路	接地回路
ground rod（＝earth electrode）	接地极	接地棒
ground tackle	锚具	錨具
group starter panel	组合起动屏	群起動盤
group velocity	群速度	群速
GRP（＝glass fiber reinforced plastics）	玻璃纤维增强塑料	玻[璃]纖[維]強化塑膠
guide bar（＝guided rod）	导向杆	導桿

英 文 名	祖国大陆名	台湾地区名
guide blade	导[向]叶[片]	導葉
guided missile boat（FAC missle）	导弹快艇	[導向]飛彈快艇
guided missile frigate（FFG）	导弹护卫舰	飛彈巡防艦
guided rod	导向杆	導桿
guideline	导向索	導索,扶手索
guide plate	导航板	導板
guide rod	导航杆	導桿
guide rope（＝guideline）	导向索	導索,扶手索
guide vane（＝guide blade）	导[向]叶[片]	導葉
guideway	导板	導板
guillotine shear（＝plane shear）	剪板机	截斷機
gumming	结胶	膠結
gun port shield	炮身窗护板	炮擋,炮圍
gun rest（＝carriage）	炮架	炮架
gun shield（＝gun port shield）	炮身窗护板	炮擋,炮圍
gusset plate	扣板	角牽板
guy	牵索	牽索,控索
guyed tower	拉索塔	張索塔（鑽油台）
gyro	陀螺仪	陀螺儀,回轉儀
gyrocompass	陀螺罗经	電羅經
gyrocompass room	陀螺罗经室	電羅經室
gyroscope（＝gyro）	陀螺仪	陀螺儀,回轉儀

H

英 文 名	祖国大陆名	台湾地区名
half beam	半梁	半梁
hand cleaning（＝handtool cleaning）	手工除锈	手工敲剷除銹
hand fire signal	手持烟火信号,手持火焰信号	手持火焰信號
hand flare（＝hand fire signal）	手持烟火信号,手持火焰信号	手持火焰信號
hand lead	测深锤	測[深]錘
hand pump	手动泵	手搖泵
handtool cleaning	手工除锈	手工敲剷除銹
hangar	机库	[飛]機庫
hanging rudder（＝spade rudder）	悬挂舵	懸舵,吊舵
harbour craft	港务船	港勤艇

英　文　名	祖国大陆名	台湾地区名
hard chine	尖舭	銳稜舭［線］,硬稜舭
hard-chine hull	折角型船艇	硬稜船體
hardenability	淬透性	可硬化性
hardener	固化剂	硬化劑
hardening capacity	淬硬性	硬化性,淬火性
hard-over angle	满舵舵角	滿舵角
hatch	舱口	艙口
hatch batten	封舱压条	艙口壓條
hatch beam (= portable hatch beam)	舱口活动横梁	艙口活動梁
hatch coaming	舱口围板	艙口緣圍
hatchcover	舱口盖	艙口蓋
hatchcover winch	舱口盖绞车	艙口蓋絞機
hatch end beam	舱口端梁	艙口端梁
hatch opening (= hatch)	舱口	艙口
hatch side cantilever beam	舱口悬臂梁	艙口側半梁
hatch side girder	舱口纵桁	艙口側縱梁
hatchway (= hatch)	舱口	艙口
hatch wedge	封舱楔	艙口楔
hauling line (= heaving line)	撇缆绳	撇纜繩,引纜繩
hawse pipe	锚链筒	錨鏈筒
hawser	缆索	大索
hawser store	帆缆间	帆纜庫,索具庫
head pump (= forepeak pump)	艏尖舱泵	艏尖艙手動泵
head reach	停船冲程	正［慣性］滑行距離
head sea	顶浪	頂浪
heat-affected zone	热影响区	熱影響區［域］（焊接）
heat balance	热平衡	熱平衡
heat exchanger	热交换器	熱交換器
heating surface	受热面	受熱面
heating surface area	换热面积	受熱面積
heat loss	热损失	熱損失
heat rate	热耗率	耗熱率
heat treatment	热处理	熱處理
heaving	垂荡	［船身］起伏
heaving line	撇缆绳	撇纜繩,引纜繩
heel (= list)	横倾	傾斜,偏斜,傾側
heeling moment	横倾力矩	傾側力矩
heel on turning	回转横倾角	回轉傾側

英 文 名	祖 国 大 陆 名	台 湾 地 区 名
heliarc welding（＝helium arc welding）	氦弧焊	氦［氩電］弧焊接
helical flow pump（＝regenerative pump）	旋涡泵	旋流泵
helicopter deck	直升机甲板	直升機甲板
helium arc welding	氦弧焊	氦［氩電］弧焊接
helm angle（＝rudder angle）	舵角	舵角
helmet	面罩	面罩
helmet shield（＝helmet）	面罩	面罩
high holding power anchor	大抓力锚	高抓著力錨
high performance craft	高性能船	高性能船
high pressure air bottle	高压空气瓶	高壓空氣瓶
high pressure boiler	高压锅炉	高壓鍋爐
high pressure feed water heater	高压［给水］加热器	高壓加熱器
high pressure heater（＝high pressure feed water heater）	高压［给水］加热器	高壓加熱器
high pressure turbine	高压透平	高壓渦輪機
high temperature corrosion	高温腐蚀	高溫腐蝕
hogging	中拱	舯拱
hoisting drum（＝drum）	卷筒	捲索鼓
hoisting winch	起落绞车	起吊絞機
hold（＝compartment）	舱	艙
hold capacity（＝cargo capacity）	货舱容积	貨艙容量
hold frame	底舱肋骨	艙肋骨
hollow blade	空心叶片	空心葉片
hollow shaft	空心轴	空心軸
hopper barge	泥驳	開底泥駁［船］
hopper capacity	泥舱容积	泥艙容量
hopper dredger	带泥舱挖泥船	斗式挖泥船
horizon indicating lamp	水平指示灯	海平面指示燈
horizontal bracket	水平肘板	水平肘板
horizontal engine	卧式内燃机	臥式引擎
horizontal flexural vibration	水平弯曲振动	水平彎曲振動
horizontal position welding	横焊	橫［向］焊［接］
horizontal welding（＝horizontal position welding）	横焊	橫［向］焊［接］
hose test	冲水试验	射水試驗
hospital ship	医院船	醫院船
hot blast	预热送风	熱風
hot cracking（＝hot tearing）	热裂	熱裂,高溫龜裂

英　文　名	祖国大陆名	台湾地区名
hot forming	热成形	加熱成型
hot shortness	热脆	熱脆性
hot spot	热节	高熱點
hot starting	热态起动	熱起動
hot tearing	热裂	熱裂,高溫龜裂
hot water boiler	热水锅炉	熱水鍋爐
hot well (= cascade tank)	热井	熱井
hover gap	飞高	裙底氣隙
hovercraft	全垫升气垫船	氣墊船
hub	桨毂	[螺槳]轂
hub cavitation	桨毂空化	轂空化
hub diameter	毂径	螺旋轂直徑
hub diameter ratio	毂径比	轂徑比
hub ratio (= hub diameter ratio)	毂径比	轂徑比
hull	船体	船體,船身,船殼
hull assembly	船体装配	
hull closures	关闭设备	關閉裝置
hull construction (= hull structure)	船体结构	船體結構
hull deflection	船体挠度	船體撓度
hull efficiency	船身效率	船殼效率
hull fittings (= outfitting)	舾装	舾裝
hull form	船体线型	船形
hull natural frequency	船体固有振动频率	船體自然頻率
hull outfit (= hull outfitting)	船装	船體舾裝
hull outfitting	船装	船體舾裝
hull-return system	船体回路系统	船體回路系統
hull steel fabrication	船体加工	
hull stiffness	船体刚度	船體剛性
hull strength	船体强度	船體強度
hull structure	船体结构	船體結構
hull torsional vibration	船体扭转振动	船體扭轉振動
hump	阻力峰	駝峰（速力曲線）
hydraulically controlled clutch	液压离合器	液壓離合器
hydraulic clutch (= hydraulically controlled clutch)	液压离合器	液壓離合器
hydraulic coupling (= oil injection coupling)	液压联轴器	液壓聯結器
hydraulic governor	液压式调速器	液壓調速器

英 文 名	祖 国 大 陆 名	台 湾 地 区 名
hydraulic machine (= hydraulic press)	液压机	液壓機
hydraulic press	液压机	液壓機
hydrofoil boat (= hydrofoil craft)	水翼艇	水翼船
hydrofoil craft	水翼艇	水翼船
hydrofoil rudder (= streamline rudder)	流线型舵	流線型舵
hydrogen controlled electrode (= low hydrogen type electrode)	低氢型焊条	低氢焊條
hydrogen embrittlement	氢脆	氢脆化
hydrogen induced cracking	氢致破裂	氢脆裂
hydrographic survey ship (= hydrographic survey vessel)	航道测量船	測量船
hydrographic survey vessel	航道测量船	測量船
hydrojet propelled ship (= water jet [propelled] boat)	喷水推进船	噴水推進船,噴射推進船
hydrological survey ship (= hydrological survey vessel)	水文调查船	
hydrological survey vessel	水文调查船	
hydrostatic curve	静水力曲线	靜水[性能]曲線圖
hydrostatic release unit	静水压力释放器	釋放裝置
hydrostatic test	水压试验	[靜]水壓[力]試驗
hysteresis	磁滞	磁滯
hysteresis loop	不稳定环	遲滯回圈,磁滯回圈

I

英 文 名	祖 国 大 陆 名	台 湾 地 区 名
IACS (= International Association of Classification Societies)	国际船级社协会	國際船級協會聯合會
IBC code (= International Code for the Construction and Equipment of Ships Carrying Dangerous Chemicals in Bulk)	国际散装运输危险化学品船舶构造与设备规则	國際散裝化學品章程
ice anchor	冰锚	冰錨
ice belt	冰带区	冰带（外板）,冰带板列
icebreaker	破冰船	破冰船
ice bunker	冰舱	冰艙
ice hold (= ice bunker)	冰舱	冰艙
ice load	冰载荷	冰負荷
ideal cycle	理想循环	理想循環

英　文　名	祖国大陆名	台湾地区名
IGC code (= International Code for the Construction and Equipment of Ships Carrying Liquefied Gases in Bulk）	国际散装运输液化气体船舶构造和设备规则	國際氣體載運船章程
IGG (= inert gas generators）	惰性气体发生装置	惰性產生器
igniter	点火器	點火器
ignition	着火,点火	點火
ignition advance angle (= spark advance angle）	点火提前角	點火提早角
ignition lag (= spark lag）	点火延迟	點火遲延
ignition plug (= igniter）	点火器	點火器
IGS (= inert gas system）	惰性气体系统	惰氣系統
IHP (= indicated horsepower）	指示功率	指示馬力
ILLC (= International Convention on Load Lines）	国际船舶载重线公约	國際載重線公約
immersion suit	救生服	浸水衣
IMO (= International Maritime Organization）	国际海事组织	國際海事組織
impact	冲击	衝擊
impingement corrosion	冲击腐蚀	衝擊腐蝕
impressed current cathodic protection	外加电流阴极保护	外加電流陰極防蝕
impulse turbine	冲动式汽轮机	衝動渦輪機
inboard turning (= inward turning）	内旋	内[向旋]轉（螺槳）
incident angle (= angle of incidence）	入射角	入射角,攻角
inclination of building slip (= slope of building berth）	船台坡度	造船台坡度
inclined engine	斜置式内燃机	斜置引擎
inclining experiment (= inclining test）	倾斜试验	傾側試驗
inclining test	倾斜试验	傾側試驗
inclusion	夹杂物	夾雜物（焊接）
incomplete penetration	①未钎透 ②未焊透	不完全穿透
indentation	压痕	壓痕
indicated horsepower （IHP）(= indicated power）	指示功率	指示馬力
indicated power	指示功率	指示馬力
indicated specific fuel consumption （ISFC）	指示油耗率	指示馬力燃油消耗率
indicated thermal efficiency	指示热效率	指示熱效率
indicator cock	示功阀	指示閥

英 文 名	祖国大陆名	台湾地区名
indicator diagram	示功图	示功圖
indicator valve（=indicator cock）	示功阀	指示閥
induction brazing	感应钎焊	［電］感應硬焊
induction heating	感应加热	［電］感應加熱
induction motor	感应电动机	感應馬達，感應電動機
inert gas	惰性气体	惰氣
inert-gas arc welding	惰性气体保护焊	惰氣［金屬］電［弧］焊， 金屬惰氣電弧焊
inert gas blower	惰性气体鼓风机	惰氣鼓風機
inert gas generators（IGG）	惰性气体发生装置	惰性産生器
inert gas system（IGS）	惰性气体系统	惰氣系統
inertial navigation system	惯性导航系统	慣性導航系統
inertia supercharging	惯性增压	慣性增壓
inflammable point（=fire point）	燃点	燃點
inflatable liferaft	气胀救生筏	充氣式救生筏
inflated appliance	充气式设备	已充氣救生設備
initial stability	初稳性	初穩度
initial survey（=original inspection）	初检查	初次檢驗
injection piston	压射冲头	柱塞
injection pressure	喷油压力	噴射壓力
injector	注水器	注射器
inland vessel	内河船	內河船，內陸水域船
inlet valve	吸气阀	進入閥
in-line engine	直列式内燃机	直列引擎，單排引擎
inner bottom longitudinal	内底纵骨	內底縱材
inner bottom plating	内底板	內底板
inner shaft	内侧轴系	內軸
inner shafting（=inner shaft）	内侧轴系	內軸
inspection after construction（=final inspection）	完工检验	完工檢查
insubmersibility	不沉性	不沉性
insulated cable	绝缘电缆	絕緣電纜
insulating material	绝缘材料	絕緣材料
insulation	绝缘层	絕緣
insulation resistance	绝缘电阻	絕緣電阻
intake-air temperature（=intake temperature）	进气温度	進氣溫度
intake pressure	进气压力	進氣壓力

英　文　名	祖国大陆名	台湾地区名
intake temperature	进气温度	進氣溫度
integrated navigation	综合导航	整合航海
interchangeability	互换性	互換性
intercontinental missile	洲际导弹	洲際彈道飛彈
intercooler	中间冷却器	中間冷卻器
interface	界面,接口	界面,接口
interference	干涉	干涉
intergranular corrosion	晶间腐蚀	粒間腐蝕
interior arrangement	舱室布置	
interlaminar shear strength	层间剪切强度	層間剪切強度
interlock arrangement (= interlocking device)	联锁机构	連鎖裝置
interlocking device	联锁机构	連鎖裝置
interlocking gear (= interlocking device)	联锁机构	連鎖裝置
intermediate frame	中间肋骨	中間肋骨
intermediate pressure turbine	中压透平	中壓渦輪機
intermediate shaft	中间轴	中間軸
intermediate shaft bearing	中间轴承	中間軸承
intermittent welding	断续焊	斷續焊法
internal combustion engine power plant	内燃机动力装置	内燃機動力設備
internal efficiency	内效率	内效率
internal stress	内应力	内應力
internal stress (= residual stress)	残余应力	殘留應力
International Association of Classification Societies (IACS)	国际船级社协会	國際船級協會聯合會
International Code for the Construction and Equipment of Ships Carrying Dangerous Chemicals in Bulk (IBC code)	国际散装运输危险化学品船舶构造与设备规则	國際散裝化學品章程
International Code for the Construction and Equipment of Ships Carrying Liquefied Gases in Bulk (IGC code)	国际散装运输液化气体船舶构造和设备规则	國際氣體載運船章程
International Convention for Safety Container (CSC)	国际集装箱安全公约	國際安全貨櫃公約
International Convention for The Prevention of Pollution from ships (MARPOL)	国际防止船舶造成污染公约	防止船舶污染國際公約
International Convention for The Safety of Life at Sea (SOLAS)	国际海上人命安全公约	海上人命安全國際公約
International Convention on Load Lines	国际船舶载重线公约	國際載重線公約

英 文 名	祖 国 大 陆 名	台 湾 地 区 名
(ILLC)		
International Convention on Tonnage Measurement of Ships	国际船舶吨位丈量公约	船舶噸位丈量國際公約
International Load Line Certificate	国际船舶载重线证书	國際載重線證書
International Maritime Organization (IMO)	国际海事组织	國際海事組織
International Oil Pollution Prevention Certificate (IOPP cert.)	国际防止油污证书	國際防止油污證書
International Pollution Prevention Certificate for The Carriage of Noxious Liquid Substance	国际防止散装运输有毒液体物质污染证书	國際載運散裝有毒液體物質防止污染證書
International Regulations for Preventing Collisions at Sea (COLREGS)	国际海上避碰规则	國際海上避碰規則
International Sewage Pollution Prevention Certificate (ISPP)	国际防止生活污水污染证书	國際防止污水污染證書
international shore connection	国际通岸接头	國際岸上接頭
International Tonnage Certificate	国际吨位证书	國際噸位證書
interstitial corrosion (= crevice corrosion)	缝隙腐蚀	間隙腐蝕,隙間腐蝕
intrinsically safe circuit	本质安全电路	本質安全線路
inward turning	内旋	内[向旋]轉（螺槳）
IOPP cert. (= International Oil Pollution Prevention Certificate)	国际防止油污证书	國際防止油污證書
iron powder type electrode	铁粉焊条	鐵粉系焊條
ISFC (= indicated specific fuel consumption)	指示油耗率	指示馬力燃油消耗率

J

英 文 名	祖 国 大 陆 名	台 湾 地 区 名
jack-up drilling unit	自升式钻井平台	升降式鑽油台
jack-up rigs (= jack-up drilling unit)	自升式钻井平台	升降式鑽油台
Jenckel rudder	差动舵	
jet dredge pump	喷射泵	噴射泵
jet ejector dredger	喷射泵挖泥船	噴射式挖泥船
jet propeller (= jet propulsion unit)	喷气推进器	噴射推進器
jet propulsion	喷气推进	噴射推進
jet propulsion ship (= water jet [propelled] boat)	喷水推进船	噴射推進船

英 文 名	祖国大陆名	台湾地区名
jet propulsion unit	喷气推进器	噴射推進器
jet pump (= jet dredge pump)	喷射泵	噴射泵
jetty	栈桥式码头	突堤碼頭
jigs and fixtures	夹具	夾具,緊固具
joining shackle	连接卸扣	連接環
jury rudder	应急舵	應急舵

K

英 文 名	祖国大陆名	台湾地区名
kedge anchor	移船锚	小錨
keel	龙骨	龍骨
keel block	龙骨墩	中心墩
keel line	龙骨线	龍骨線
king post	吊杆柱	主柱
knee (= bracket)	肘板	肘板,腋板
knee plate (= bracket)	肘板	肘板,腋板

L

英 文 名	祖国大陆名	台湾地区名
labyrinth packing	迷宫式汽封	迷宫迫緊,曲徑填封
lacing wire	拉筋	系索
lagging	保温层	[隔熱]襯套
landing craft	登陆艇	登陸艇
lap joint	搭接接头	搭接
lapping	研磨	研光
lap weld (= lap welding)	搭接焊	搭焊
lap welding	搭接焊	搭焊
Laser velocimeter	激光测速仪	雷射測速儀
LASH (= lighter aboard ship)	载驳船	駁船搬運船,子母船,浮貨櫃船
lateral underwater area	水下侧面积	水下侧[向]面積
launching	下水	下水
launching barge	下水驳	下水駁船
launching cradle	下水车	下水船架
launching way	滑道	下水滑道,下水台
launching weight	下水重量	下水重量

英 文 名	祖 国 大 陆 名	台 湾 地 区 名
launch retrieval apparatus	吊放回收装置	吊放回收裝置
layer	焊层	焊層
layer corrosion	层状腐蚀	層狀腐蝕
laying depth	埋设深度	埋設深度
laying down（＝laying-off of hull lines）	型线放样	放樣
laying-off of hull lines	型线放样	放樣
laying out	划线	劃線
layout（＝laying out）	划线	劃線
L-bar（＝l-section）	L 型钢	角鋼
L-drive	L 型传动	L 型傳動
leading block	舵链导轮	導滑車
leading edge	导边	導緣,前緣
leakage	逸流	逸流,泄漏
leakage current	泄漏电流	漏電流
leakage loss	漏汽损失	泄漏損失
leak-off	泄漏	漏出
left hand model	左开式	左開式
left-hand revolving engine unit	左转机组	左轉機組
left-hand turning	左旋	左旋
leg	焊脚	焊腳（填角焊）
length	船长	船長
length between perpendiculars	垂线间长	垂標間距
length breadth ratio	长宽比	長寬比
length depth ratio	长深比	長深比
length of chain cable	链节	錨鏈長度
length of entrance	进流段长	艏入水段長
length of run	去流段长	艉出水段長
length overall（LOA）	总长	全長
level gauge	液位计	液位計
lifeboat	救生艇	救生艇
lifebuoy	救生圈	救生圈
lifebuoy self-activating smoke signal	救生圈自发烟雾信号	救生圈自發煙霧信號
life float	救生浮	［兩面用硬式］救生筏
lifejacket	救生衣	救生衣
lifejacket chest	救生衣柜	救生衣櫃
lifejacket light	救生衣灯	救生衣燈
life line	安全索	救生索,攀手索
life-line	救生索	救生索

英　文　名	祖国大陆名	台湾地区名
liferaft	救生筏	救生筏
life rope	放艇安全索	放艇安全索
life-saving appliance	①救生设备 ②救生属具	①救生設備 ②救生屬具
lift effect damping	升力阻尼	升力阻尼
lift fan	垫升风扇	墊升風機
lift hatchcover	升降式舱口盖	升降式艙口蓋
lifting analysis	吊装分析	吊装分析
lifting and mounting complete superstructure	上层建筑整体吊装	
lifting beam	吊梁	吊梁
lifting hook	艇吊钩	吊鈎
lifting magnet	电磁吸盘	起重磁鐵
lifting pontoon	打捞浮筒	打捞浮筒
lift out piston	吊缸	吊缸
light alloy ship	轻合金船	輕合金船
light boat	灯船	燈標船
light buoy crane	航标起重机	燈浮標起重機
light derrick boom	轻型吊杆	輕型吊桿
light diesel oil transfer pump	轻柴油输送泵	輕柴油輸送泵
light draft	空载吃水	輕載吃水
lightened floor	轻型肋板	減輕孔肋板
lightening hole	减轻孔	減輕孔
lighter（＝barge）	驳船	駁船
lighter aboard ship（LASH）（＝barge carrier）	载驳船	子母船, 浮貨櫃船
lightning arrester（＝surge arrester）	避雷器	避雷器
light weight	空船重量	輕載
light weight distribution	空船重量分布	空船重量分佈
lignumvitae	铁梨木	鐵梨木
lime titania type electrode	钛钙型焊条	鹼基氧化鈦系焊條
limit switch	限位开关	極限開關
line of cut	切割线	切割線
liner cable laying machine	履带式布缆机	履帶式布纜機
lines（＝lines plan）	型线图	線型[圖]
lines fairing	型线光顺	線型整順
lines plan	型线图	線型[圖]
line winder	盘线装置	捲線機

英　文　名	祖国大陆名	台湾地区名
lining	衬里	内襯,襯料,襯層
liquefied gas carrier	液化气体船	液化氣體船
liquefied natural gas carrier（＝LNG carrier）	液化天然气船	液化天然氣船
liquefied natural gas tank	液化天然气舱	液化天然氣艙
liquefied petroleum gas carrier（＝LPG carrier）	液化石油气船	液化石油氣船
liquefied petroleum gas tank	液化石油气舱	液化石油氣艙
liquid-filled pressure vacuum breaker	充液式压力真空安全装置	
liquid tank	液舱	液艙
list	横倾	傾斜,偏斜,傾側
live fish carrier	活鱼运输船	活魚運輸船,活魚船
living quarter（＝accommodation）	起居舱室	住艙
LNG carrier	液化天然气船	液化天然氣船
LOA（＝length overall）	总长	全長
load	负荷,负载	負荷,負載
load carrying capacity	承载能力	負荷能量
load curve	载荷曲线	①負荷曲線 ②載重曲線
load draught（＝loaded draft）	满载吃水	滿載吃水
loaded draft	满载吃水	滿載吃水
load indicator	装载仪	負荷指示器
load line	载重线	載重[水]線
load line mark	载重线标志	載重線標誌
load shedding	分级卸载	卸載
load test	负荷试验	負荷試驗
load waterline	满载水线	載重水線
load waterline length	满载水线长	
local control	就地控制	現場控制
localized corrosion	局部腐蚀	局部腐蝕
local strength	局部强度	局部強度
local vibration	局部振动	局部振動
locating	定位	定位
locking bar	封舱锁条	鎖緊柄
lock-in lock-out submersible	闸式潜水器	閘式潛水器
lock nut	保险螺母	並緊螺帽
locus of metacenters	稳心曲线	定傾中心曲線圖

英 文 名	祖 国 大 陆 名	台 湾 地 区 名
lofting	[船体]放样	放樣
log	计程仪	計程儀
log carrier (= timber carrier)	运木船	木材運輸船,運木船
long crested waves	长峰波	長峰波
longitudinal	纵骨	縱材
longitudinal bending	总纵弯曲	縱[向]彎曲
longitudinal bending stress (= normal str- ess due to longitudinal bending mo- ment)	总纵弯曲正应力	縱向彎應力
longitudinal bulkhead	纵舱壁	縱[向]艙壁
longitudinal center of buoyancy	浮心纵向坐标	縱向浮[力中]心
longitudinal center of floatation	漂心纵向坐标	縱向浮[力中]心
longitudinal center of gravity	重心纵向坐标	縱向重心
longitudinal framing system	纵骨架式	縱肋系統
longitudinal girder	纵桁	縱桁
longitudinal metacenter	纵稳心	縱定傾中心
longitudinal metacentric height	纵重稳距	縱定傾中心高
longitudinal section in center place	中纵剖面	中心縱剖面
longitudinal stability	纵稳性	縱向穩度
longitudinal strength	总纵强度	縱向強度
longitudinal strength member	纵向强力构件	縱向強度構件
longitudinal strength test	总纵强度试验	縱向強度試驗
longitudinal vibration	纵向振动	縱向振動
longitudinal vibration of shafting (= axial vibration of shafting)	轴系纵向振动	軸系縱向振動
loop scavenging	回流扫力	環流驅氣,環流掃氣
Loran A	罗兰 A	羅遠 A 導航系統
Loran C	罗兰 C	羅遠 C 導航系統
Loran inertial navigation system	罗兰 - 惯性导航系统	羅遠 - 慣性導航系統
lost waterplane area	损失水线面面积	損失水線面面積
lower dead center (= bottom dead center)	下止点	下死點
lower dead point	下死点	下死點
lower decks	下甲板	下甲板
lower mast	桅柱	下桅
lower tumbler	下导轮	下導輪
low hydrogen type electrode	低氢型焊条	低氫焊條
low-magnetic steel	低磁钢	非磁性鋼
low pressure air bottle	低压空气瓶	低壓空氣瓶

英　文　名	祖国大陆名	台湾地区名
low pressure steam generator	低压蒸汽发生器	低壓蒸汽產生器
low pressure turbine	低压透平	低壓渦輪機
low sea suction valve	低位通海阀	低位海水吸入閥
low speed wind tunnel	低速风洞	低速風洞
low temperature corrosion	低温腐蚀	低溫腐蝕
LPG carrier	液化石油气船	液化石油氣船
L-section	L 型钢	角鋼
lubricating oil batch purification	滑油间歇净化	滑油間歇淨化
lubricating oil cooler	滑油冷却器	［潤］滑油冷卻器
lubricating oil drain system	滑油泄放系统	滑油泄放系統
lubricating oil drain tank	滑油泄放柜	滑油泄放櫃
lubricating oil filter	滑油滤器	［潤］滑油過濾器
lubricating oil heater	滑油加热器	滑油加熱器
lubricating oil pump	滑油泵	［潤］滑油泵
lubricating oil purifying system	滑油净化系统	滑油淨化系統
lubricating oil system	滑油系统	滑油系統
lubricating oil tank	滑油舱	［潤］滑油櫃
lubricating oil transfer pump	滑油输送泵	滑油輸送泵
lug connection	直接连接	直接連接
luggage room	行李舱	行李間
lugger（=pearl boat）	采珍渔船	采珠船
lugless joining shackle	连接链环	連接鏈環
luminance	亮度	亮度

M

英　文　名	祖国大陆名	台湾地区名
machine finish allowance（=machining allowance）	加工余量	機制裕度
machinery fitting	机装	輪機艤裝
machinery space of category A	A 类机器处所	甲種機艙空間
machining	机械加工	機械加工
machining allowance	加工余量	機制裕度
magnetic blow	磁偏吹	磁吹
magnetic compass	磁罗经	磁羅經
magnetic field	磁场	磁場
magnetic flux	磁通	磁通［量］
magnetic particle testing	磁粉检测	磁粉探傷檢查

英　文　名	祖国大陆名	台湾地区名
magnetic pole	磁极	磁極
magneto hydrodynamic propulsion plant	磁流体动力推进装置	磁流體動力推進裝置
magnetostrictive transducer	磁致伸缩换能器	磁致伸縮換能器
mail room	邮件舱	郵件艙
main air bottle	主空气瓶	主空氣瓶
main ballast tank	主压载水舱	主壓載艙
main boiler	主锅炉	主鍋爐
main cable	主干电缆	主電纜
main diesel engine	主柴油机	主柴油機
main electric power source (= main source of electrical power)	主电源	主電源
main engine	主机	主機
main engine bed (= main engine foundation)	主机基座	主機座, 主機台
main engine foundation	主机基座	主機座, 主機台
main feed system	主给水系统	主給水系統
main frame	主肋骨	主肋骨
main gas turbine	主燃气轮机	主燃氣渦輪機
main gas turbine set	主燃气轮机组	主燃氣渦輪機組
main generating set	主发电机组	主發電機組
main hook load	主钩起重量	主鉤起重量
main hull	主船体	主船殼
main shaft	[汽轮机]主轴	主軸
main source of electrical power	主电源	主電源
main steam turbine	主汽轮机	主汽輪機
main steam turbine set	主汽轮机组	主[蒸]汽[渦]輪機組
main steering gear	主操舵装置	主操舵裝置
main stop valve	主汽阀	主停止閥
main switchboard	主配电板	主配電盤
maintainability	维修性	維護性
maintenance	维修	維護, 保養
make-up feed water	补给水	補充給水
male union	旋入管接头	外螺紋活管套節
malleable cast iron	可锻铸铁	展性鑄鐵
maneuverability test	操纵性试验	操縱性試驗
maneuvering light	操纵信号灯	操船信號燈
man hole	人孔	人孔
manhole cover	人孔盖	人孔蓋

英　文　名	祖　国　大　陆　名	台　湾　地　区　名
manifold valve	阀箱	歧管阀
manipulator	焊接操作机	自動操控機
manned submersible	载人潜水器	有人潛水器
manual control	人工控制	人工控制,手動控制
manual emergency anchoring test	人力应急起锚试验	人力緊急起錨試驗
manual emergency steering test	人力应急操舵试验	人力緊急操舵試驗
manual fire alarm sounder	手动火灾警报器,手动火灾警报装置	手動火災警報系統
manual fire alarm system (= manual fire alarm sounder)	手动火灾警报器,手动火灾警报装置	手動火災警報系統
manual override system	人工越控装置	人工越控裝置
manual welding	手工焊	人工焊接
margin line	限界线	邊際線
margin plate	内底边板	二重底緣板
marine accumulator batteries	船用蓄电池组	
marine air bottle	船用气瓶	
marine automatic fire alarm system	船用火灾自动报警装置	
marine auxiliary machinery	船舶辅机	船舶輔機
marine boiler	船用锅炉	船用鍋爐
marine diesel engine	船用柴油机	船用柴油機
marine electrical equipment	船舶电气设备	船舶電氣設備
marine electrical power plant	船舶电站	
marine engine	船用内燃机	船用引擎
marine fire fighting system	船舶消防装置	
marine fire proof panel	船用防火板	船用防火板
marine furniture	船用家具	船用家具
marine galley equipment	船用厨房设备	船用廚房設備
marine gas turbine	船用燃气轮机	船用燃氣渦輪機
marine hydropneumatic components and units	船用液压气动元件	
marine incinerator	船用焚烧炉	船用焚化爐
marine paint	船舶涂料	船舶塗料
marine plywood	船用胶合板	船用合板
marine pollution	海洋污染	海洋污染
marine power plant	船舶动力装置	
marine public address system	船用广播设备	船用廣播系統
marine pump	船用泵	船用泵
marine refrigerating plant	船舶制冷装置	

英 文 名	祖 国 大 陆 名	台 湾 地 区 名
marine sanitary fixtures	船用卫生设备	船用衛生設施
marine sanitation device（MSD）（＝marine sanitary fixtures）	船用卫生设备	船用衛生設施
marine sewage treatment system	船用生活污水处理装置	船用［生活］污水處理系統
marine steam turbine	船用汽轮机	船用［蒸］汽［渦］輪機
marine storage batteries（＝marine accumulator batteries）	船用蓄电池组	
marine transformer	船用变压器	船用變壓器
marine-type fan	船用风机	船用風機
marine valves	船用阀门	船用閥門
maritime engineering（＝ship engineering）	船舶工程	
marking of hull parts	［船体零件］号料	
MARPOL（＝International Convention for The Prevention of Pollution from ships）	国际防止船舶造成污染公约	防止船舶污染國際公約
mask（＝helmet）	面罩	面罩
masking	遮蔽	遮蔽
mass-balance weight	配重	配重
mast	桅	桅
mast and rigging	桅设备	桅檣設備
mast coating	桅杆涂料	桅桿塗料
master compass	主罗经	主羅經
master controller	主令控制器	主控制器
master switch	主令电器	主控開關,總開關
mast head light	桅灯	桅燈
mast light（＝mast head light）	桅灯	桅燈
mat jack-up drilling platform（＝mat jack-up drilling unit）	沉垫自升式钻井平台	
mat jack-up drilling unit	沉垫自升式钻井平台	
mathematical lofting	数学放样	數學放樣
maximum blade width（＝maximum width of blade）	最大叶宽	最大葉寬
maximum blade width ratio	最大叶宽比	最大葉寬比
maximum combustion pressure	最高燃烧压力	最高燃燒壓力
maximum continuous rating（MCR）	最大持续功率	連續最大出力,額定最大連續出力
maximum section	最大横剖面	最大橫剖面

英 文 名	祖 国 大 陆 名	台 湾 地 区 名
maximum speed	最高转速	最高速[率]
maximum transverse section coefficient	最大横剖面系数	
maximum width of blade	最大叶宽	最大葉寬
MCR (=maximum continuous rating)	最大持续功率	連續最大出力,額定最大連續出力
MCT (=multi-cables transit)	积木式电缆填料盒	多管穿線板
mean blade width	平均叶宽	平均葉寬
mean blade width ratio	平均叶宽比	平均葉寬比
mean draft	平均吃水	平均吃水
mean piston speed	活塞平均速度	活塞平均速度
mean repair time	平均修复时间	平均修復時間
mean sea level	平均海平面	平均水位
measuring instruments	计量器具	量測儀器
measuring section	测试段	測試段
mechanical draft	机械通风	機械通風
mechanical efficiency	机械效率	機械效率
mechanical governor	机械式调速器	機械[式]調速機
mechanical load	机械负荷	機械負荷
mechanical noise	机械噪声	機械噪音
mechanical rust removal	机械除锈	機械除銹
mechanical ventilation (=mechanical draft)	机械通风	機械通風
member	构件	構件
membrane wall	膜式水冷壁	薄膜壁
merchant ship	商船	商船
metacenter	稳心	定傾中心
metacentric diagram (=locus of metacenters)	稳心曲线	定傾中心曲線圖
metacentric height	重稳距	定傾[中心]高度
metacentric radius	稳心半径	定傾半徑
metal coating	金属覆盖层	金屬覆蓋層
metal hot dipping	金属热浸镀	金屬熱浸鍍
metal spraying	金属喷镀	金屬噴敷
midship section	中横剖面	舯[横]剖面
midship section coefficient	中横剖面系数	舯剖面係數
midstation plane	中站面	中站面
mill scale	热轧氧化皮	軋鋼鱗片
mine	水雷	水雷

英　文　名	祖国大陆名	台湾地区名
mine anchor	雷锚	雷錨
mine compensating tank	水雷式代换水舱	水雷補重櫃（潛艇）
minehunter	猎雷舰艇	獵雷艦, 獵雷艇
mine hunting	猎雷	獵雷
minelayer	布雷舰艇	佈雷艦, 佈雷艇
minesweeper	扫雷舰艇	掃雷艇
mining dredger	采矿船	
missile range instrumentation ship	导弹卫星跟踪测量船	飛彈衛星跟蹤測量船
mitre caisson	人字式坞门	人字式塢門
mixed cycle	混合循环	混合循環
mixer	混合器	混合器, 調合機
mixing chamber	混合室	混合室
mixing condenser	混合式凝汽器	混合冷凝器
mobile auxiliary gas turbine	移动式辅燃气轮机	移動式輔燃氣［渦］輪機
mobile drilling platform（= mobile drilling unit）	移动式钻井平台	移動式鑽探平台, 可動式鑽油台
mobile drilling unit	移动式钻井平台	移動式鑽探平台, 可動式鑽油台
mobile platform（= mobile unit）	移动式平台	移動式平台
mobile unit	移动式平台	移動式平台
mock-up	样箱	模型打樣
module	模块	模組, 模件
moisture resistant insulating material	耐潮绝缘材料	抗濕絕緣材料
mold	铸型	鑄模
molded breadth	型宽	型寬, 模寬
molded depth	型深	型深, 模深
molded displacement	型排水量	型排水量, 模排水量
molded draft	型吃水	型吃水量, 模吃水
molded hull surface	船体型表面	船［體］型面
molded lines	型线	型線, 模線
molded volume	型排水体积	型排水體積
molding	造型	造模翻砂, 成型
molding machine	造型机	造模機, 塑造機
molten pool	熔池	熔［化］池（焊接）
moment of inertia	转动惯量	慣性矩
moment of inertia of midship section	舯剖面惯性矩	舯剖面慣性矩
mono-hull ship	单体船	單體船

英 文 名	祖 国 大 陆 名	台 湾 地 区 名
moored condition	系泊状态	系泊狀態
mooring anchor	固定锚	系留錨,碇泊錨
mooring arrangement test	系泊设备试验	系泊裝置試驗
mooring buoy	系船浮筒	系泊浮筒
mooring cleat	带缆羊角	帶纜羊角
mooring equipment	系泊设备	系泊設備
mooring fittings	系缆具	系纜具
mooring line	系缆索	系泊索
mooring pipe	导缆孔	[系船]導索管
mooring swivel	系泊旋转接头	系船轉環
mooring trial	系泊试验	系泊試車
Morse signal light	莫氏信号灯	莫氏信號燈
mother ship of submersible	潜水器母船	潛水器母船
mother-ship with fishing dory	母子式渔船	
motion compensation equipment	运动补偿设备	運動補償設備
motor	电动机	電動機,馬達
motor lifeboat	机动救生艇	動力救生艇,馬達救生艇
motor sailer (= power-sail ship)	机帆船	機帆船
motor ship (= diesel ship)	柴油机船	柴油機船,內燃機船
motor siren	电动膜片式气笛	電動氣笛
moving blade	动叶[片]	[轉]動葉片
MSD (= marine sanitation device)	船用卫生设备	船用衛生設施
mud box	泥箱	泥箱
multi-bladed rudder	多叶舵	多葉舵
multi-buoy mooring system	多浮筒系泊系统	多浮筒系泊系統
multi-cables transit（MCT）	积木式电缆填料盒	多管穿線板
multi-core cable	多芯电缆	多芯電纜
multi-decked ship	多甲板船	多甲板船
multi-hulled craft (= multi-hulled ship)	多体船	多體船
multi-hulled ship	多体船	多體船
multi-layer welding	多层焊	多層焊接
multi-pintle rudder	多支承舵	多舵針舵
multiple-pile driver tower	多能打桩架	多能打樁架
multipressure turbine	多压式汽轮机	多壓渦輪機
multipurpose cargo ship	多用途货船	多用途貨船
multi-stages feed heating	多级预热给水	多級預熱給水
Munk moment	孟克力矩	孟克力矩

N

英　文　名	祖国大陆名	台湾地区名
naked hull (= bare hull)	裸船体	裸船殼
name plate	识别板	銘牌
natural draft	自然通风	自然通風
natural draught (= natural draft)	自然通风	自然通風
naval ship	舰艇,军舰	軍船
navigation	导航	航海
navigational aid	导航设备	助航設施
navigational draft	外形吃水	外型吃水
navigation bridge (= wheel house)	驾驶室	駕駛室,〔操〕舵房
navigation deck	驾驶甲板	航海甲板,橋樓甲板
navigation light (= running light)	航行灯	航行燈
navigation light indicator	航行灯控制器	航行燈指示器
navigation parameter	导航参数	導航參數
navigation satellite	导航卫星	導航衛星
navigation signal equipment	航行信号设备	航行信號設備
navigation sonar	导航声呐	導航聲納
NDI (= nondestructive inspection)	无损检验	非破壞檢查,非破壞試驗
NDT (= nondestructive testing)	无损检验	非破壞檢查,非破壞試驗
nearest land	最近陆地	最近陸地
neck bearing	下舵承	下舵承
negative hull-return D. C. single system (= negative hull-return D. C. single-wire system)	利用船体作负极回路的直流单线系统	直流單線船體負極回路系統
negative hull-return D. C. single-wire system	利用船体作负极回路的直流单线系统	直流單線船體負極回路系統
net carrying pipe	送网管	送網管
net drum	卷网机	捲網機
net hauler (= net winch)	起网机	起網機
net hauling system	起网机组	起網系統
net horsepower (= net power)	净功率	淨馬力
net layer	布网船	布網船

英 文 名	祖 国 大 陆 名	台 湾 地 区 名
net power	净功率	淨馬力
net shifter	理网机	理網機
net tonnage	净吨位	淨噸位
net winch	起网机	起網機
neutral flame	中性焰	中性火焰
nil-ductility transition temperature	无塑性转变温度	無延性轉變溫度
node	节点	節點
noise	噪声	噪音
no-load operation	空载运行,空负荷运行	無負荷運轉
no-load running（=no-load operation）	空载运行,空负荷运行	無負荷運轉
no-load test	空负荷试验	空載試驗,無負載試驗
no-load voltage	空载电压	無負載電壓
non-combustible material	不燃材料	不燃材料
nondestructive inspection（NDI）	无损检验	非破壞檢查,非破壞試驗
nondestructive testing（NDT）（=nondestructive inspection）	无损检验	非破壞檢查,非破壞試驗
non-magnetic steel（=low-magnetic steel）	低磁钢	非磁性鋼
non-metallic inclusions	非金属夹杂	非金屬夾雜物
non-penetrated hole	盲孔	盲孔
non-powered ship	非机动船,非自航船	無動力船
non-stationary cavities	不稳定空泡	不穩定空泡
non-watertight bulkhead	非水密舱壁	非水密艙壁
non-watertight door	非水密门	非水密門
norm	定额	定額
normal condition	正常状态	正常狀況
normal stress due to longitudinal bending moment	总纵弯曲正应力	縱向彎應力
not under command light	失控灯	[船]操縱失靈號燈
nozzle	喷嘴	噴嘴,噴口
nozzle box	喷嘴室	噴嘴箱
nuclear [-powered] ship	核动力船	核能動力船
nuclear power plant	核动力装置	動子動力設備
nuclear submarine	核潜艇	核子潛艇
nugget	熔核	焊塊

O

英　文　名	祖国大陆名	台湾地区名
oar-propelled lifeboat	划桨救生艇	
OBO（=ore/bulk/oil carrier）	油、散、矿船	礦砂散裝貨與油兼用船
ocean going ship	远洋船	遠洋船舶
ocean going vessel（=ocean going ship）	远洋船	遠洋船舶
oceanographic research ship（=oceano-graphic research vessel）	海洋调查船	海洋研究船
oceanographic research vessel	海洋调查船	海洋研究船
oceanography survey ship（=oceano-graphic research vessel）	海洋调查船	海洋調查船
octave	倍频程	倍頻帶
offset	错移	偏位
offsets	型值	船線座標
offsets table（=table of offsets）	型值表	船線坐標表
offshore platform（=platform）	[海洋]平台	海域平台
offshore unit（=platform）	[海洋]平台	海域平台
oil carrier（=oil tanker）	油船	油輪
oil cooler	冷油器	油冷[卻]器
oil cooling	油冷	油冷
oil discharge monitoring and control system（=oil discharge monitoring system）	排油监控系统	泄油監控系統
oil discharge monitoring system	排油监控系统	泄油監控系統
oil filtering equipment	滤油设备	濾油設備
oil injection coupling	液压联轴器	液壓聯結器
oil purifier	滤油器	淨油機
oil sealing	油封	油封
oil skimmer	浮油回收船	撈油船
oil storage tanker	储油船	儲油船
oil sump	油底壳	油池,油槽
oil tank	油箱	油櫃,油艙
oil tanker	油船	油輪
oil test	油压试验	驗油,油驗
oil tight bulkhead	油密舱壁	油密艙壁
oil-water interface detector	油水界面探测器	油水界面計

英 文 名	祖 国 大 陆 名	台 湾 地 区 名
oily mixture	油性混合物	含油混合物
oily water separator	油污水分离装置	油水分離器
oily-water separating equipment	油水分离设备	油水分離裝置
open feed system	开式给水系统	開式給水系統
open lifeboat	开敞式救生艇	敞式救生艇
open-up examination	拆验	拆檢
open water propeller efficiency	螺旋桨敞水效率	螺槳單獨效率
opposed-piston engine	对动活塞式内燃机	對冲活塞引擎
optimization	优化	最適化
ordered rudder angle	指令舵角	指令舵角
ore/bulk/oil carrier（OBO）	油、散、矿船	礦砂散裝貨與油兼用船
ore carrier	矿砂船	礦砂船
ore/oil carrier	矿油船	礦砂與油兼用船
original inspection	初检查	初次檢驗
outboard engine	舷外挂机	舷外機
outer shaft（= outer shafting）	外侧轴系	外側軸
outer shafting	外侧轴系	外側軸
outfit of deck and accommodation	舾装设备	
outfitting	舾装	舾裝
outreach（= boom outreach）	舷外跨距	伸出［舷外］距離
outrigger	桅肩	桅上固定具
outward turning	外旋	外旋
overboard discharge valve	舷外排出阀	舷側排泄閥
over current	过电流	過［量］電流
overflow pipe	溢流管	溢流管
overhang	调伸长度	懸伸
overhead position	仰焊位置	仰姿
over heat（= overheating）	过热	過熱
overheating	过热	過熱
overlapping joint	搭接结点	搭接接頭
overlapping propeller（= overlap propeller）	重叠螺旋桨	重疊型螺槳
overlap propeller	重叠螺旋桨	重疊型螺槳
overload	①超负荷 ②过负载	①過［量］負荷 ②過載，超載
overload test	超负荷试验	過負荷試驗
overshoot angle	超越角	過沖角
overspeed	超速	超速

英　文　名	祖国大陆名	台湾地区名
overspeed protection device	超速保护装置	超速保護設施
overtorque protection device	超扭保护装置	超扭〔矩〕保護設施
oxiding flame (= oxidizing flame)	氧化焰	氧化焰
oxidizing flame	氧化焰	氧化焰
oxy-acetylene flame	氧乙炔焰	氧乙炔焰
oxy-acetylene welding	氧乙炔焊	氧乙炔焊接
oxygen breathing apparatus	氧气呼吸器	氧氣呼吸器
oxygen lance cutting	氧矛切割	氧吹管切割
oxy-hydrogen welding	氢氧焊	氫氧焊接

P

英　文　名	祖国大陆名	台湾地区名
pack	包装件	包裝件
package	包装	包裝
package (= pack)	包装件	包裝件
packing (= package)	包装	包裝
paddle wheel	明轮	明輪
paint store	油漆间	油漆庫
painting	涂装,涂料施工	塗裝,塗漆
pallet	托盘	托〔貨〕板
Panama canal tonnage	巴拿马运河吨位	巴拿馬運河噸位
Panama Canal Tonnage Certificate	巴拿马运河吨位证书	巴拿馬運河噸位證書
Panama chock	巴拿马运河导缆孔	巴拿馬運河導索器
panel	板列	板列,列板
panting arrangement	强胸结构	抗拍結構
panting beam	强胸横梁	抗拍梁
parallel middle body	平行中体	平行舯體
parallel operation	并联运行	並聯運轉（電）
parent metal (= base metal)	母材	母材
partial bulkhead	局部舱壁	部分艙壁
partial load	部分负荷	部分負荷
partially enclosed lifeboat	部分封闭救生艇	部分圍蔽救生艇
partially under hung rudder	半悬舵	半懸舵
parting agent	分型剂	脫模劑
partition bulkhead	轻舱壁	隔間艙壁
passage	通道	通道
passenger cabin	客舱	客艙

英　文　名	祖国大陆名	台湾地区名
passenger-cargo ship	客货船	客貨輪
passenger ship	客船	客船
Passenger Ship Safety Certificate	客船安全证书	客船安全證書
passenger vessel（＝passenger ship）	客船	客船
pattern	模样	模型
payload	①有酬载荷 ②有效载荷	①酬载 ②有效負載
peak frame	尖舱肋骨	尖艙肋骨
pearl boat	采珍渔船	采珠船
pedestal roller	导向滚轮	台式滾輪
peeling	起皮	剝離
penetration	熔深	熔解深度
perforated distribution plate	多孔板	多孔板
perforated plate（＝perforated distribution plate）	多孔板	多孔板
performance test	性能试验	性能試驗
periscope	潜望镜	潛望鏡
permanent current	连续电流	連續電流
permanent magnet	永久磁铁	永久磁鐵
permissible length	许可舱长	許可長度
petrol engine（＝gasoline engine）	汽油机	汽油機
phase advancer	进相机	進相機（電）
phase convertor	变相机	換相機
pickling	①强浸蚀 ②酸洗	①浸洗 ②酸洗
piercing（＝punching）	冲孔	沖孔,穿孔
pig iron	生铁	生鐵
pilot hoist	引航员升降装置	引水員升降機
pilot ladder	引航员软梯	引水[人]梯
pilot valve	错油门	導閥
pinhole	针孔	針孔,銷孔
pipe band	管卡	管夾,管固定帶
pipe hanger	管子吊架	管吊架
piston	活塞	活塞
pitch	螺距	螺距
pitch angle	螺距角	螺距角
pitching	纵摇	縱搖
pitch ratio	螺距比	節圓直徑比
pitting	点蚀	斑蝕,孔蝕

英　文　名	祖国大陆名	台湾地区名
pitting corrosion（=pitting）	点蚀	斑蚀,孔蚀
pivoting point	枢心	轉向軸心
planar motion mechanism test（PMM test）	平面运动机构试验	平面運動機構試驗
plane bulkhead	平面舱壁	平面艙壁
plane shear	剪板机	截斷機
planing boat	滑行艇	滑行[快]艇,水上快艇,滑航艇
plasticity	塑性	塑性
plate keel	平板龙骨	平板龍骨
platform	[海洋]平台	海域平台
pleasure craft	游艇	遊艇
plough-in	埋艏	艏潛,埋艏
plug welding	塞焊	塞孔焊接
plug-hole	堵孔	塞孔
plummer block（=intermediate shaft bearing）	中间轴承	中間軸承
plunger（=injection piston）	压射冲头	柱塞
plunger pump	柱塞泵	柱塞泵
PMM test（=planar motion mechanism test）	平面运动机构试验	平面運動機構試驗
pneumatic hammer	空气锤	[空]氣錘
polar expedition ship	极地考察船	極地考察船
polarization	极化	極化
pole and line fishing boat	竿钓渔船	竿釣漁船
pontoon	①趸船 ②浮箱	①躉船 ②浮箱
pontoon floating dock	浮箱式浮船坞	
pontoon hatchcover	箱形舱口盖	箱形艙蓋
poop	艉楼	艉樓
popping pressure	起座压力	泄氣壓力（安全閥）
port of registry	船籍港	船籍港
portable fire extinguisher	手提式灭火器	輕便滅火器
portable hatch beam	舱口活动横梁	艙口活動梁
portable pump	可移式泵	移動式泵
post heating（=postheat）	后热	後熱（焊接）
postheat	后热	後熱（焊接）
postweld heat treatment	焊后热处理	焊接後熱處理
pounding	捶击	拍底

英 文 名	祖 国 大 陆 名	台 湾 地 区 名
powder fire extinguisher	干粉灭火器	乾粉滅火器
power	功率	功率
power-driven ship	机动船,自航船	機動船,動力船,自航船
power-driven vessel（＝power-driven ship）	机动船,自航船	機動船,動力船,自航船
power-sail ship	机帆船	機帆船
power system	动力系统	動力系統
power take-off（PTO）	辅助功率输出装置,分出功率输出装置	輔助功率輸出裝置
prechamber	预燃[燃烧]室	預燃室
pre-combustion chamber（＝prechamber）	预燃[燃烧]室	預燃室
pre-cooler	预冷器	預冷器
preheat	预热	預熱
preheater	预热器	預熱器
preheating（＝preheat）	预热	預熱
pre-ignition	早燃	先期點火
pre-outfitting	预舾装	預舾裝
pressure drop	汽水阻力	壓力降
pressure hull	耐压壳体	壓力殼（潛艇）
pressure ratio	[增]压比	壓力比
pressure tank	压力箱	壓力水櫃
pressure test	压力试验	壓力試驗
pressure welding	压焊	壓接
primary air	①一次空气 ②一次风	①主空氣 ②初級空氣
primary member	①主要构件 ②重要构件	主構件
primer	底漆	底漆
priming	①汽水共腾 ②涂底漆	①汽水共腾 ②刷底漆
principal dimensions	主尺度	主要尺寸
principal particulars	主要要素	主要特徵,主要項目
prismatic coefficient	棱形系数	棱形係數,[縱向]棱塊係數
product carrier	成品油船	油品船
production storage tanker	生产储油船	
production test ship	生产测试船	
profile	外廓线	船體側面[圖]
projected area	投影面积	投影面積
projected area ratio	投影面积比	投影面積比

英　文　名	祖国大陆名	台湾地区名
projection welding	凸焊	凸出焊接
promenade deck	游步甲板	散步甲板
propeller	推进器	推进器,螺桨
propeller boss	轴毂	螺[桨]毂
propeller cap	毂帽	螺桨帽
propeller fitting force	螺旋桨推入力	安装螺桨壓擠力
propeller plan	螺旋桨平面	螺桨平面
propeller post	推进器柱	螺桨柱
propeller pull-up distance	螺旋桨推入量	安装螺桨壓擠量
propeller race	螺旋桨尾流	螺桨尾流,[螺桨]舻流
propeller racing	飞车	螺桨空车,飛车
propeller shaft (= screw shaft)	螺旋桨轴,舻轴	舻[管]轴
propeller shaft bracket	舻轴架	舻轴架,螺桨支架
propeller strut (= propeller shaft bracket)	舻轴架	舻轴架,螺桨支架
propeller vessel (= screw ship)	螺旋桨船	螺桨船
propelling plant	推进装置	推进設備
propulsion electric machine	推进电机	
propulsion units	推进机组	推进装置
propulsive coefficient	推进系数	推进係數
propulsive efficiency	推进效率	推进效率
propulsive fan	推进风机	推进風機
propulsor (= propeller)	推进器	推进器
protection current density	保护电流密度	保護電流密度
protective clothing	防护服	防護衣
protective coating	保护性覆盖层	保護性塗層
protective cover	防护罩	護罩
protective potential	保护电位	保護電位
provision store	粮食库	糧食庫,給養艙
PTO (= power take-off)	辅助功率输出装置,分出功率输出装置	輔助功率輸出裝置
pulsating cavity	脉动空泡	脈動空泡
pulse repetition frequency	脉冲重复频率	脈衝複現頻率
pump room	泵舱	泵室
punch	凸模	沖頭
punching	冲孔	沖孔,穿孔
purifier	净化器	淨化器
push boat (= pusher)	[顶]推船	推[型拖]船
pusher	[顶]推船	推[型拖]船

Q

英 文 名	祖 国 大 陆 名	台 湾 地 区 名
quadrant	舵扇	舵柄弧
quality control	质量控制	品[質]管[制]
quality factor	品质因数	品質因數
quarantine vessel	检疫船	檢疫船
quartering sea	艉斜浪	艉側浪
quench cracking	淬火冷却开裂	淬火裂痕
quenching	淬[火]冷[却]	急冷,淬火
quenching and tempering	调质	淬火及回火
quenching crack (= quench cracking)	淬火冷却开裂	淬火裂痕
quenching stresses	淬火冷却应力	淬火應力
quenching temperature	淬火冷却起始温度	淬火溫度
quick acting hatchcover (= quick-closing hatchcover)	速闭舱盖	快速開閉艙蓋
quick-closing hatchcover	速闭舱盖	快速開閉艙蓋
quick diving	速潜	緊急下潛

R

英 文 名	祖 国 大 陆 名	台 湾 地 区 名
racing boat	赛艇	競賽艇
racon	雷康	雷[達示]標
radar beacon	雷达信标	雷[達示]標
radar cross section (RCS)	雷达截面	雷達截面積
radar mast	雷达桅	雷達桅
radial davit	转出式吊艇架	旋臂吊架
radial engine	星形内燃机	星形引擎
radial flow turbine	辐流式汽轮机	徑向流渦輪機
radio beacon	无线电信标	無線電示標
radio frequency interference (= radio interference)	射频干扰	無線電頻率干擾
radiographic inspection	射线透照检查	放射線檢查
radio interference	射频干扰	無線電頻率干擾
radio navigation	无线电导航	無線電導航

英　文　名	祖国大陆名	台湾地区名
radio room	报务室	無線電室
raft	筏	筏
rail（=railing）	栏杆	欄桿
railing	栏杆	欄桿
rake	纵斜	後傾（螺葉）
rake angle（=rake angle of propeller）	螺旋桨纵倾角	後傾角（螺槳）
rake angle of propeller	螺旋桨纵倾角	後傾角（螺槳）
ramp	跳板	跳板,著陸板
ramp door	跳板门	登陸舌門,登陸跳板
ram-wing craft	冲翼艇	沖翼船
range	射程	射程
RAS（=replenishment at sea）	海上补给	海上整補
rated output	额定[输出]功率	額定出力
rated power	额定功率	額定功率
rat guard	防鼠板	防鼠板
RCS（=radar cross section）	雷达截面	雷達截面積
reach	复向期	回復期
reaction rudder	反应舵	反動式舵
reaction turbine	反动式汽轮机	反動式渦輪機
reactor room	反应堆舱	[核子]反應爐艙
receiving transducer	水听器,接收换能器	
rectangular window	矩形窗	
rectifier	整流器	整流器
recuperator	表面式回热器	複熱器
reduction gear	减速齿轮箱	減速裝置
reel barge	卷筒铺管驳	捲盤駁船（海洋工程）
refractoriness	耐火度	耐火度,耐火性
refrigerant	冷处理剂	冷凍劑,冷媒
refrigerated cargo hold	冷藏货舱	冷藏貨艙
refrigerated carrier（=refrigerator ship）	冷藏船	冷藏船,冷凍船
refrigerating chamber	冷藏库	冷藏庫,冷凍庫
refrigerating medium pump	制冷剂泵	冷媒泵
refrigeration container	消耗式冷剂冷藏集装箱	冷凍貨櫃,冷藏貨櫃
refrigerator ship	冷藏船	冷藏船,冷凍船
regenerative air heater（=rotary air heater）	回转式空气预热器	再生式空氣加熱器
regenerative heat exchanger	再生式热交换器	再生式熱交換器
regenerative pump	旋涡泵	旋流泵

英　文　名	祖国大陆名	台湾地区名
regenerative turbine	回热式汽轮机	再生式渦輪機
regenerator	回热器	再生器
register	调风器	調氣器
register of shipping（＝classification soci-ety）	船级社	船級協會,船級機構
regular wave	规则波	規則波
reheat	再热	再熱
reheat factor	重热系数	再熱因數
reheater	再热器	再熱器
reinforced concrete ship	钢筋混凝土船	鋼筋混凝土船
reinforced concrete vessel（＝reinforced concrete ship）	钢筋混凝土船	鋼筋混凝土船
reinforcement	①加强层 ②余高	①加強材 ②焊縫凸量
relative rotative efficiency	相对旋转效率	［相］對轉［動］效率
relay	继电器	繼電器,替續器
release device（＝hydrostatic release unit）	静水压力释放器	釋放裝置
reluctance	磁阻	磁阻
remote control	远距离控制	遙控
remote-controlled mine	遥控水雷	遙控水雷
repair ship	修理船	修理艦
repair welding	补焊	補焊
replenishment at sea（RAS）	海上补给	海上整補
rescue bell	救生钟	救生鍾
rescue boat	救助艇	救難艇
rescue ship	救助船	救難船
reserve buoyancy	储备浮力	預［留］浮力
reserve feedwater tank	补水储存舱	備用給水櫃
residual magnetism	剩磁	剩磁
residual oil standard discharge connection	残油类标准排放接头	殘油排泄標準接頭
residual resistance	剩余阻力	剩餘阻力
residual stress	残余应力	殘留應力
residuary resistance（＝residual resist-ance）	剩余阻力	剩餘阻力
resilient shaft coupling	弹性联轴器	彈性聯軸器
resistance brazing	电阻钎焊	電阻硬焊
resistance coefficient	阻力系数	阻力係數
resistance dynamometer	船模阻力仪	阻力儀

英　文　名	祖国大陆名	台湾地区名
resistance of an earthed conductor	接地电阻	接地電阻
resistance test	阻力试验	阻力試驗
resistance test in waves	波浪中阻力试验	波浪中阻力試驗
resistance welding	电阻焊	電阻焊接
resonance	共振	共振
resonance speed (= critical speed)	临界转速	臨界速率,共振轉速
response spectrum	响应谱	反應[波]譜
restoring lever (= righting lever)	复原力臂	扶正力臂
restoring moment (= righting moment)	复原力矩	扶正力矩,復原力矩,回復力矩
restricted maneuver light	操纵限制灯	操縱限制燈
restricted water effect	狭航道效应	局限水域效應
return stroke	回程	回行衝程
reverberatory furnace	反射炉	反射爐
reversed frame	内底横骨	副肋骨
reverse osmosis desalination device	反渗透海水淡化装置	逆滲透[海水淡化]裝置
reverse osmosis device (= reverse osmosis desalination device)	反渗透海水淡化装置	逆滲透[海水淡化]裝置
reverse spiral test	逆螺线试验	逆蝸旋試驗
reversing gear	换向机构	逆轉裝置
reversing test	换向试验	逆轉試驗
reversing time	换向时间	換向時間
rigging	索具	索具
rigging screw	螺旋扣	緊索螺釘
right hand model	右开式	右開式
right-hand turning	右旋	右旋
righting lever	复原力臂	扶正力臂
righting moment	复原力矩	扶正力矩,復原力矩,回復力矩
rigid liferaft	刚性救生筏	硬式救生筏
ring plate	眼环	環板
riser	上升管,立管	升導管
riser	冒口	冒口（鑄造）
riveting	铆接	鉚接
Ro/Ro access equipment	滚装通道设备	滾裝通道設備
Ro/Ro ship (= roll on-roll off ship)	滚装船	滾裝船,駛上駛下船,轆轆船

英 文 名	祖国大陆名	台湾地区名
rock cutter suction dredger	铰吸式挖石船	鉸刀吸入式挖泥船
rocket launcher	火箭发射装置	火箭發射器
rocket parachute flare (= rocket parachute flare signal)	火箭降落伞火焰信号	火箭式降落傘照明彈
rocket parachute flare signal	火箭降落伞火焰信号	火箭式降落傘照明彈
rocket star signal	红星火箭	火箭信號
rocket signal (= rocket star signal)	红星火箭	火箭信號
roll-angle	横摇角	橫搖角
roller	轧辊	滾子,輥子
roller fairlead	滚轮导缆器	滾子導索器
roller fair-leader (= roller fairlead)	滚轮导缆器	滾子導索器
roller painting	辊涂	滾輪式塗漆
rolling	①横摇 ②轧制	①[船身]橫搖 ②軋
rolling angle (= roll-angle)	横摇角	橫搖角
rolling chock (= bilge keel)	舭龙骨	舭龍骨
rolling damping	横摇阻尼	橫搖阻尼
rolling experiment (= rolling test)	摇摆试验	橫搖試驗
rolling hatchcover	滚动式舱口盖	滾動[式]艙口蓋
rolling moment (= rolling torsional moment)	横摇扭矩	橫搖力矩
rolling test	摇摆试验	橫搖試驗
rolling torsional moment	横摇扭矩	橫搖力矩
roll on-roll off ship	滚装船	滾裝船,駛上駛下船,轆轆船
roll stowing hatchcover	滚卷式舱口盖	捲動[式]艙口蓋
root cavitation	叶根空化	葉根空化
root crack	焊根裂纹	根部龜裂
root face	钝边	根面（焊接）
root gap	根部间隙	根隙（焊接）
root pass	根部焊道	初層焊道
root radius	根部半径	根[部]半徑
root thickness	根厚	葉根厚度
root vortex	根涡	葉根渦旋
rope ladder	软梯	軟梯,繩梯
ropeless linkage	无缆系结装置	無纜系結裝置
rope reel	卷纲机	繩索捲盤,捲索盤
rope stopper	掣索器	制索器
rope storage reel	缆索卷车	纜索捲車

英 文 名	祖国大陆名	台湾地区名
rotary air heater	回转式空气预热器	再生式空氣加熱器
rotary convertor	旋转变流机	旋轉式換流機
rotating arm basin	旋臂水池	旋臂水槽
rotating-arm test	旋臂试验	［船模］强制回旋試驗
rotating cylinder rudder	转柱舵	轉柱舵
rotating field (= rotational magnetic field)	旋转磁场	旋轉磁場
rotational magnetic field	旋转磁场	旋轉磁場
rotor	转子	轉子
rough water resistance (= rough-sea re-sistance)	汹涛阻力	洶濤阻力
roughness resistance	粗糙度阻力	粗面阻力
rough-sea resistance	汹涛阻力	洶濤阻力
round-bilge hull	圆舭艇型	圓舭型船體
rubberized dredge pump	胶衬泥泵	膠襯泥泵
rudder	舵	舵
rudder and steering gear	舵设备	舵裝置
rudder angle	舵角	舵角
rudder angle indicator (= electric rudder angle indicator)	舵角指示器	舵角指示器
rudder area ratio	舵面积比	舵面積比
rudder arm	舵臂	舵臂
rudder axle	舵轴	舵軸
rudder balance area ratio (= coefficient of balance of rudder)	舵平衡比	舵平衡面積比
rudder bearing	舵承	舵軸承
rudder blade	舵叶	舵葉
rudder brake	舵掣	舵靭,舵制動器
rudder carrier	上舵承	舵承上部,上舵承
rudder coupling	舵杆接头	舵頸接頭
rudder frame	舵构架	舵肋
rudder gear (= rudder and steering gear)	舵设备	舵裝置
rudder gudgeon	舵钮	舵針承
rudder head	舵头	上部舵桿,舵頭材
rudder horn	挂舵臂	半懸舵承架
rudder main stock	主舵杆	舵葉主構體
rudder pintle	舵销	舵針
rudder plate	舵板	舵板
rudder post	舵柱	舵柱

英　文　名	祖国大陆名	台湾地区名
rudder pressure	舵压力	舵壓
rudder stock	舵杆	舵桿
rudder stop	舵角限位器	舵止器, 制舵器
rudder stopper (= rudder stop)	舵角限位器	舵止器, 制舵器
rudder stuffing box (= stuffing box)	舵杆填料函	舵填料函
rudder torque	舵杆扭矩	舵轉矩
rudder yoke	横舵柄	舵軛, 横舵柄
rules for classification and construction of ships	船舶入级与建造规范	船舶入級與建造規範
run	去流段	艉出水段
runner boat (= racing boat)	赛艇	競賽艇
running-in	磨合	適配運轉
running light	航行灯	航行燈
running light indicator (= navigation light indicator)	航行灯控制器	航行燈指示器
running trial	运转试验	運轉試驗
rust preventive oil	防锈油	防銹油

S

英　文　名	祖国大陆名	台湾地区名
sacrificial anode	牺牲阳极	陽極耗蝕
sacrificial anode cathodic protection (= galvanic anode protection)	牺牲阳极阴极保护	犧牲陽極陰極保護
saddle chamber	鞍形舱	鞍形艙
safe-light	安全灯	安全燈
safety	安全开关	安全開關
safety lamp (= safe-light)	安全灯	安全燈
safety switch (= safety)	安全开关	安全開關
safety valve	安全阀	安全閥
safe working load (SWL)	吊杆安全工作负荷	安全工作負荷, 安全使用負荷
sagging	中垂	舯垂[现象]
sail	指挥室围壳	[潛艇]指揮室
sailer	帆船	帆船
sailing boat	帆艇	帆船
SALM (= single anchor leg mooring)	单锚腿系泊装置	單腳錨泊
SALS (= single anchor leg storage)	单锚腿储油装置	

英　文　名	祖国大陆名	台湾地区名
salvage barge	救捞驳	救難艇
salvage boat (= salvage barge)	救捞驳	救難艇
salvage pump	救助泵	救難泵
salvage ship	打捞船	救難船
sampan	舢板	舢板
sand blast (= sand blasting)	喷砂	噴砂［除銹］
sand blasting	喷砂	噴砂［除銹］
sand block (= block with sand box)	砂箱墩	砂箱墩,砂墩
sanitary pump	卫生泵	衛生水泵
sanitary unit	卫生单元	衛生單元
satellite navigation	卫星导航	衛星航法
SBT (= segregated ballast tank)	专用压载水舱	隔離壓載水艙
scale effect	尺度效应	尺度效應
scale lofting	比例放样	
scale ratio	缩尺比	尺度比
scantling draft	结构吃水	強度［計算］吃水
scavenging air receiver	扫气箱	驅氣接收器
scavenging efficiency	扫气效率	驅氣效率
scoop circulating water system	自流式循环水系统	自然式循環水系統
scouring	淘空	海泥流失（鑽油井）
scout (= scout ship)	侦察船	偵查艦
scout ship	侦察船	偵查艦
scraping of bearing	轴承刮削	軸承刮削
scraping of propeller boss	螺旋桨锥孔研配	螺轂孔研配
screen	屏蔽层,遮障	簾幕,隔板,隔屏
screen bulkhead (= partition bulkhead)	轻舱壁	屏隔艙壁
screw conveyor	螺旋输送机	螺旋輸送機
screw press	螺旋压力机	螺［旋］壓機
screw propeller	螺旋桨	螺［旋］槳
screw shaft	螺旋桨轴,艉轴	艉［管］轴
screw ship	螺旋桨船	螺槳船
scupper	漏水口	排水孔
sea anchor (= floating anchor)	浮锚	海錨
sea chest cleaning valve	吹除阀	海底門清除閥
seacock	通海旋塞	海底旋塞
sea condition	海况	海況,海面狀態
sea direction	浪向	浪向
sea-going ship	海船	海船,海輪,遠洋船

英 文 名	祖国大陆名	台湾地区名
sea suction valve	通海阀	通海閥,海水吸入閥
sea trial	航行试验	［海上］試航
sea water desalting plant	海水淡化装置	海水淡化裝置
sea water distillate plant (＝sea water distillation plant)	海水蒸馏装置	海水蒸餾裝置
sea water distillation plant	海水蒸馏装置	海水蒸餾裝置
sea water filter	海水滤器	海水濾器
sea water pump	海水泵	海水泵
seakeeping performance (＝seakeeping qualities)	耐波性	耐波性能
seakeeping qualities	耐波性	耐海性
seakeeping tank	耐波性试验水池	耐海性試驗水槽
seakeeping test	耐波性试验	耐海性試驗
sealing	封闭	封閉
sealing run (＝back weld)	封底焊道	背焊道
sealing welding	封底焊	防漏焊
seal weld (＝sealing welding)	封底焊	防漏焊
seam arrangement	板缝排列	
search coil	感应线圈	探索線圈
seat	座板	底座
seaworthiness	适航性	適航性
secondary member	次要构件	次［要］構件
second deck	第二甲板	第二［層］甲板
sectional dock	分体式浮船坞	分段浮塢
sectional floating drydock (＝sectional dock)	分体式浮船坞	分段浮塢
sectional method of hull construction	分段建造法	分段建造法
section assembly (＝fabrication)	分段装配	［船體］分段裝配
section board	区配电板	分段配電板,區配電板
section outfitting	分段舾装	
segregated ballast	专用压载水	隔離壓艙水
segregated ballast tank (SBT)	专用压载水舱	隔離壓載水艙
seiner	围网渔船	圍網漁船,巾著網漁船
selective corrosion	选择性腐蚀	選擇腐蝕
self-emptying installation	抽舱排泥装置	抽艙排泥裝置
self-igniting buoy light	自亮浮灯	［救生圈］自燃燈
self polishing copolymer antifouling paint	自抛光防污涂料	自抛光防污漆,自磨型防污漆

英 文 名	祖 国 大 陆 名	台 湾 地 区 名
self propelled dredger	自航挖泥船	自航式挖泥船
self-propelled vessel (' = power-driven ship)	机动船, 自航船	機動船, 動力船, 自航船
self propelling dredger (= self propelled dredger)	自航挖泥船	自航式挖泥船
self-propulsion factor	自航因子	自推因子
self-propulsion test	自航试验	自推試驗 (船模試驗)
self-regulation	自调	自動調節
semi-balanced rudder (= partially under hung rudder)	半悬舵	半平衡舵
semi-closed feed water system	半闭式给水系统	半閉式給水系統
semi-dock building berth	半坞式船台	半塢式船台
semi-submerged ship	半潜船	半潛式船
semi-submersible drilling platform (= semi-submersible drilling unit)	半潜式钻井平台	半潛式鑽探平台
semi-submersible drilling unit	半潜式钻井平台	半潛式鑽探平台
sensitivity	灵敏度	靈敏度, 敏感度
sensor	传感器	感測器
service boat (= service craft)	基地勤务船	港勤艇
service craft	基地勤务船	港勤艇
service speed	服务航速	航海船速, 營運船速
servomotor	油动机	伺服馬達
set-back	翘度	[螺槳]後傾
sewage	生活污水	穢水, [生活]污水
sewage holding tank	生活污水储存柜	生活污水儲存櫃
sewage pump	粪便泵	穢水泵, 污水泵
sewage treatment plant	生活污水处理装置	穢水處理裝置
sewage water pump	污水泵	污水泵
sextant	六分仪	六分儀
shackle (= length of chain cable)	链节	錨鏈一節
shaft alignment (= centering for shafting)	轴系校中	軸線校準, 軸線校中
shaft cut off test	减轴运转试验	减軸運轉試驗
shafting	轴系	軸系
shafting alignment (= centering for shafting)	轴系校中	軸線校準, 軸線校中
shafting brake	轴系制动器	軸遊轉防止裝置
shafting efficiency (= transmission efficiency of shafting)	轴系效率	軸系效率

英 文 名	祖 国 大 陆 名	台 湾 地 区 名
shafting-grounding device	轴系接地装置	軸接地裝置
shafting vibration	轴系振动	軸系振動
shaft liner	轴套	軸套,軸襯
shaft-locking device (= shafting brake)	轴系制动器	軸遊轉防止裝置
shaft tunnel	轴隧	軸道
shallow water effect	浅水效应	淺水效應
sheathed deck	覆材甲板	包板甲板,被覆甲板
shed	货棚	風雨棚
sheer	舷弧	舷弧
sheer strake	舷顶列板	舷側厚板列
sheet cavitation	片状空化	片狀空化
sheet metal	板料	金屬薄板,板金
shell and tube heat exchanger	管壳式热交换器	殼管式熱交換器
shell plate	外板	船殼板,外板
shell plate development	外板展开	外板展開
sheradizing	渗锌	滲鋅
shielded metal arc welding (SMAW)	手弧焊	金屬被覆電弧焊
shift winch	移动绞机	移動絞機
ship	船[舶]	船,艦
shipboard power cables	船用电力电缆	船用電力電纜
shipboard radio-frequency cables	船用射频电缆	船用射頻電纜
shipboard telecommunication cables	船用通信电缆	船用通信電纜
ship bottom coating (= ship bottom paint)	船底涂料	船底塗料,船底漆
ship bottom paint	船底涂料	船底塗料,船底漆
shipbreaking	拆船	拆船
shipbuilding electrode	船用焊条	船用焊條
ship conversion	船舶改装	船舶改裝
ship-directional quay	引船驳岸	引船駁岸
ship elevator (= shiplift)	升船机	升船機
ship engineering	船舶工程	
shiplift	升船机	升船機
ship maneuverability	船舶操纵性	船舶操縱性
ship model	船模	船模
ship noise	船舶噪声	船舶噪音
ship oscillation	船舶摇荡	船舶搖蕩
shipping route	航线	航線
ship propulsion	船舶推进	船舶推進
ship resistance	船舶阻力	船舶阻力

英 文 名	祖国大陆名	台湾地区名
ship resistance and performance	船舶快速性	船舶阻力性能
ship side pipe	船舷接管	船舷接管
ship side valve	舷侧阀	舷侧閥
ship's light	号灯	號燈
ship speed	航速	航速
ship system	船舶系统	
ship vibration	船体振动	船體振動
shop primer	车间底漆	防銹底漆
shore connecting plant	排泥管接岸装置	排泥管接岸裝置
shore connection	甲板通岸接头	通岸接頭
shore connection box	岸电箱	岸電接線盒
shore connection cables	岸电电缆	岸電電纜
short circuit	短路	短路
short-circuit current	短路电流	短路電流
short crested waves	短峰波	短峰波
shotblaster (= shot blaster chamber)	喷丸机	噴珠除銹機
shot blaster chamber	喷丸机	噴珠除銹機
shot blasting	喷丸	噴[射鋼]珠除銹,珠[粒噴]擊
shoulder	肩	水線肩部
shrinkage	①收缩②铸件线收缩率	①收縮②收縮量
shrinkage allowance	收缩余量	收縮裕度
shrinkage fit (= shrinkage fitting)	热装	收縮配合,紅套
shrinkage fitting	热装	收縮配合,紅套
shrinkage rule	缩尺	收縮尺
shrouded propeller (= ducted propeller)	导管推进器	導罩螺槳
shroud ring	转动套环	押環（汽旋機）
shuttle tanker	穿梭油船	穿梭油輪,短程往返油輪
side frame (= side framing)	船侧骨架	側肋骨
side framing	船侧骨架	側肋骨
side girder	旁桁材	側縱梁
side hopper barge	侧开泥驳	側漏斗型駁船
side keel block	边墩	邊墩
side keelson	旁内龙骨	側內龍骨
side lamp (= side light)	舷灯	舷燈
side launching	横向下水	橫向下水
side light	舷灯	舷燈

英 文 名	祖国大陆名	台湾地区名
side light (= side scuttle)	舷窗	舷[侧圆]窗
side longitudinal	舷侧肋骨	舷侧縱材
side plating	舷侧外板	船侧外板列
side port	舷门	舷侧艙口
side power roller	舷边动力滚柱	舷侧动力滚子
side rolling hatchcover	侧移式舱口盖	侧移式艙口蓋
side scuttle	舷窗	舷[侧圆]窗
sideslip	横移	侧滑
side slipway	横向滑道	横向滑道
side stringer	舷侧纵桁	侧加强肋
side thrust device	侧向推力装置	侧推装置
side transverse	舷侧竖桁	侧深横肋
side waller (= sidewall hovercraft)	侧壁气垫船	侧壁式氣垫船
sidewall hovercraft	侧壁气垫船	侧壁式氣垫船
SIGMA welding (= inert-gas arc welding)	惰性气体保护焊	惰氣[金屬]電[弧]焊, 金屬惰氣電弧焊
signal	信号	信號
signal flag	号旗	信號旗
signal line	信号绳	信號繩
signal mast	信号桅	信號桅
significant wave height	有义波高	有義波高
significant wave period	有效周期	有義波周期
silencer	消声器	消音器
silo	圆筒仓	[散裝貨]储艙
simplex rudder	舵轴舵	[舵]軸兼[舵]柱型舵
single acting pump	单作用泵	單動泵
single anchor leg mooring (SALM)	单锚腿系泊装置	單脚锚泊
single anchor leg storage (SALS)	单锚腿储油装置	
single bottom	单底	單底
single-cylinder engine	单缸[内燃]机	單缸機
single-decked ship	单甲板船	單層甲板船
single-decked vessel (= single-decked ship)	单甲板船	單層甲板船
single-pass welding	单道焊	單道焊接
single plate rudder	平板舵	平板舵,單板舵
single point mooring unit	单点系泊装置	單點系泊装置
single pull hatchcover	滚翻式舱口盖	單拉[式]艙口蓋
single unit floating dock	整体式浮船坞	

英　文　名	祖国大陆名	台湾地区名
siren (= air horn)	气笛	汽笛,氣笛;號笛
skeg	艉鳍	艉鰭
skew angle	侧斜角	[螺葉]歪斜角度
skew back	侧斜	[螺]葉歪斜
skirt	围裙	氣裙,圍裙
skirt depth	围裙高度	氣裙升高度
skirting rise (= skirt depth)	围裙高度	氣裙升高度
skirt plate	裙板	裙板
skylight	天窗	天窗
slack meter	电缆松紧指示器	電纜鬆緊計
slamming	砰击	波擊
slamming load	砰击载荷	波擊負荷
sleeve coupling	套筒联轴器	套筒聯結器
sleeve expansion joint (= telescopic joint)	[松套]伸缩接头	[套筒]伸縮接頭
sleeve joint	套管连接	套筒接合
slewing angle	吊杆偏角	吊桿旋角
slewing winch	回转绞车	[吊桿]回旋絞車
slide	滑块	滑塊
sliding block (= slide)	滑块	滑塊
sling	绳扣	吊索,吊鏈
slip hook	弹钩	滑鉤
slip ring	集电环	滑環
slipstream (= propeller race)	螺旋桨尾流	[螺槳]艉流
slipway cradle (= slipway turn cradle)	滑道摇架	滑道托架
slipway turn cradle	滑道摇架	滑道托架
slipway turntable	滑道转盘	滑道轉盤
slope of building berth	船台坡度	造船台坡度
slope of wave surface	波倾角	波面斜率,波面傾角
slope of ways (= slope of building berth)	船台坡度	船台斜度
sloping bulkhead	斜舱壁	斜艙壁
slop tank	污液舱	汙油[水]櫃
sloshing	晃荡	沖激
sludge pump	油渣泵	汙油泵
sludge tank	污泥柜	汙油[泥]櫃
small craft	小艇	小艇
small waterplane area twin hull (SWATH)	小水线面双体船	小水線面雙體船
SMAW (= shielded metal arc welding)	手弧焊	金屬被覆電弧焊
Smith correction	波浪水动压力修正	史密斯修正波效應

英 文 名	祖国大陆名	台湾地区名
Smith effect	史密斯效应	史密斯效應
smoke detector	感烟式探测器	探煙器
smokescreen	烟幕	煙幕
smoke signal	信号烟雾	煙[霧信]號
smoke tube	烟管	煙管
snip end	切斜端	切角端
snorkel	柴油机通气管工作装置	[潛艇]通氣管裝置
snort（=snorkel）	柴油机通气管工作装置	[潛艇]通氣管裝置
soil discharging facility	排泥设备	排泥設施
SOLAS（=International Convention for The Safety of Life at Sea）	国际海上人命安全公约	海上人命安全國際公約
solder	软钎料	軟焊料,焊錫
soldering	软钎焊	軟焊
sole piece	艉柱底骨	舵跟材
solid ballast	固体压载	固體壓載
solid buoyancy material	固体浮力材料	固體浮材
solid dielectric cables	实芯绝缘电缆	實芯絕緣電纜
solid floor	实肋板	實體肋板
solid shaft	实心轴	實心軸
solution heat treatment	固溶热处理	溶液熱處理
solvent	溶剂	溶劑
sonar cable winch	声呐电缆绞车	聲納電纜絞車
sonar dome coating（=sonar dome paint）	声呐导流罩涂料	聲納罩塗料
sonar dome paint	声呐导流罩涂料	聲納罩塗料
soot blower	吹灰器	吹灰器
sounding（=depth sounder）	测深	測深
sounding lead（=hand lead）	测深锤	測深[鉛]錘
sounding pipe	测深管	測深管
sounding rod	测深尺	測深標尺
sound lead（=hand lead）	测深锤	測深[鉛]錘
sound powered telephone	声力电话机	聲力電話
sound pressure	声压	聲壓
sound signal instrument	音响信号器具	音響信號器具
space	舱室	艙室
spacer	隔叶块	隔片,間隔件
space tracking ship	航天测量船	
spade rudder	悬挂舵	懸舵,吊舵
span	跨距	跨距

英 文 名	祖国大陆名	台湾地区名
span ringging	千斤索具	跨索具
span rope (=topping lift)	千斤索	跨索
span winch	变幅绞车	跨索絞機
spanwire winch (=span winch)	变幅绞车	跨索絞機
spark advance angle	点火提前角	點火提早角
spark arrester	火星熄灭器	火花防止器
spark ignition engine	点燃式内燃机	引燃機
spark lag	点火延迟	點火遲延
spatter	飞溅	焊濺物
special area	特殊区域	特別海域
specific air consumption	空气消耗率	空氣消耗率,單位耗氣量
specific fuel consumption	[燃]油[消]耗率	單位耗油量
specific lubricating oil consumption	滑油消耗率	滑油消耗率,單位耗滑油量
specific power	比功[率]	比馬力,比功率
specific steam consumption (=steam rate)	汽耗率	汽耗率
specimen	试件,试样	試件,試樣
spectacle blank flange (=spectacle blind flange)	双环盲板法兰	眼鏡型盲凸緣
spectacle blind flange	双环盲板法兰	眼鏡型盲凸緣
speed changer	转速变换器	變速器
speed control device	调速装置	速度控制器
speed controller (=speed control device)	调速装置	速度控制器
speed over the ground log	绝对计程仪	絕對計程儀
speed regulation characteristic	转速调整特性	調速特性
speed through the water log	相对计程仪	相對計程儀
speed trial	航速试验	速率試車
spiral duct	螺旋风管	螺旋風管,螺旋導管
spiral test	螺线试验	蝸旋試驗
splash zone	飞溅区	飛濺區
splash zone corrosion	飞溅区腐蚀	飛濺區腐蝕
split hull	对开船体	開體船
splittrail (=split-type trailing suction dredger)	对开耙吸挖泥船	對開式耙吸船
split-type trailing suction dredger	对开耙吸挖泥船	對開式耙吸船
sponson deck	舷伸甲板	張出甲板,舷外平台

英　文　名	祖国大陆名	台湾地区名
spot welding	点焊	點焊
spray gun	喷枪	噴槍
spraying	喷涂	噴塗
spraying tube	洒水短管	灑水短管
spray painting（＝spraying）	喷涂	噴塗
spray resistance	飞溅阻力	濺水阻力
spray transfer	喷射过渡	噴灑狀傳遞
spread anchoring system	散射锚泊系统	分散錨泊
spreader	集装箱吊具	貨櫃吊具
spread moored drilling ship	散射系泊定位钻井船	分散系泊定位鑽探船
springing	波激振动	船體波振
sprinkler	喷洒装置	噴灑裝置,撒佈器
sprinkler system	喷淋设备	噴水系統
spud	定位桩	柱錨[挖泥船]
spud carriage	定位桩台车	定位樁台車
spud driving and pulling condition	插拔桩状态	插拔樁狀態
spud for antislip	抗滑桩	抗滑樁
spud gantry	定位桩架	定位樁架
spud leg	桩腿	樁腿
square netter	敷网渔船	敷網漁船
squid angling boat	鱿鱼钓渔船	鱿[魚]釣船
squid angling machine	鱿鱼钓机	鱿魚釣機
stability	稳性	穩度
stability against sliding	抗滑稳定性	抗滑穩定性
stabilized gyrocompass	平台罗经	平台羅經
stabilizer	稳燃器	穩定器
stabilizer control unit	减摇装置控制设备	穩度器控制裝置
stabilizer fin（＝stabilizing fin）	稳定板	穩定鰭
stabilizing fin	稳定板	穩定鰭
stage efficiency	级效率	級效率
staggered intermittent fillet weld	交错断续角焊缝	交錯間斷式填角焊
stairway and passage way arrangement	梯道布置	梯道與通道佈置
stalling rudder angle	临界舵角	臨界舵角
standard discharge connection	标准排放接头	標準排泄接頭
standard displacement	标准排水量	標準排水量（軍艦）
stand-by generating set	备用发电机组	備用發電機組
stand by pump	备用泵	備用泵
standing wave	驻波	駐波

英 文 名	祖国大陆名	台湾地区名
star chain	星形链	星形鏈
starboard	右舷	右舷
starter	起动器	起動器
starting air distributor	起动空气分配器	起動空氣分配器
starting air system	起动空气系统	起動空氣系統
starting current	起动[过程]电流	起動電流
starting device	起动装置	起動設施
starting pressure	起动压力	起動壓力
starting test	起动试验	起動試驗
starting torque	起动[过程]转矩	起動轉矩,起動扭矩
statical stability	静稳性	靜穩度
statical stability curve	静稳性曲线	靜穩度曲線
static balance test	静平衡试验	靜平衡試驗
stationary blade	静叶[片]	定子輪葉
stationary engine	固定式内燃机	固定引擎
stationary wave (= standing wave)	驻波	駐波
stator blade (= stationary blade)	静叶[片]	定子輪葉
statutory survey	法定检验	法定檢驗
steady cavities	定常空泡	穩定空泡
steady state	稳态	穩態
steady turning diameter	回转直径	迴旋直徑
steady turning period	稳定回转阶段	穩定回旋階段
steam boiler	蒸汽锅炉	蒸汽鍋爐
steam chest	蒸汽室	汽櫃
steam consumption	汽耗量	蒸汽消耗量
steam distribution device	配汽机构	配汽機構
steam engine power plant	蒸汽机动力装置	蒸汽機動力設備
steamer (= steam ship)	蒸汽机船	汽船,輪船
steam fire extinguishing system	蒸汽灭火系统	蒸汽窒火系統
steam power plant	蒸汽动力装置	蒸汽動力裝置
steam rate	汽耗率	汽耗率
steam seal (= steam seal gland)	汽封	汽封
steam seal gland	汽封	汽封
steam ship	蒸汽机船	汽船,輪船
steam smothering system (= steam fire extinguishing system)	蒸汽灭火系统	蒸汽窒火系統
steam strainer	蒸汽滤器	濾汽器
steam turbine	汽轮机	蒸汽渦輪機

英　文　名	祖 国 大 陆 名	台 湾 地 区 名
steam turbine ship	汽轮机船	汽船
steam whistle	蒸汽警笛	汽笛
steel ship	钢船	鋼［殼］船
steering apparatus（＝steering gear）	操舵装置	操舵裝置
steering chain	操舵链	舵鏈
steering engine room	舵机舱	舵機艙
steering gear	①操舵装置 ②舵机	操舵裝置
steering gear room（＝steering engine room）	舵机舱	舵機艙
steering light	操舵目标灯	拖航燈
steering rod	操舵拉杆	操舵桿
steering shaft（＝steering shafting）	操舵轴	操舵軸
steering shafting	操舵轴	操舵軸
steering stand	操舵台	操舵台
steering wheel	操舵轮	舵輪
steering wire	操舵索	操舵鋼索
stem	艏柱	艏柱,艏材
stem structure	艏部结构	艏結構
stern anchor	艉锚	艉錨
stern barrel	尾滚筒	艉滾筒
stern construction（＝stern structure）	艉部结构	艉段結構
stern-engined ship	艉机型船	艉機型船,艉機艙船
stern light	艉灯	艉燈
sternline winch	艉锚绞车	艉錨絞車
stern port	艉门	
stern position winch（＝sternline winch）	艉锚绞车	艉錨絞車
stern post	艉柱	艉柱
stern ramp	艉滑道	艉斜道,艉坡道
stern roll（＝stern barrel）	尾滚筒	艉滾筒
stern shaft（＝screw shaft）	螺旋桨轴,艉轴	艉［管］軸
stern shaft seal（＝stern shaft sealing）	艉管密封装置	艉軸封
stern shaft sealing	艉管密封装置	艉軸封
stern sheave	艉滑轮	艉槽輪（布纜船）
stern structure	艉部结构	艉段結構
stern transom plate	艉封板	艉封板,艉橫板
stern trawler	尾拖渔船	艉拖網漁船
stern tube	艉管	艉軸管
stern-tube bearing	艉管轴承	艉軸套軸承

英　文　名	祖国大陆名	台湾地区名
stern-tube nut	艉管螺母	艉軸管環首螺帽
stern-tube stuffing box	艉管填料函	艉軸管填料函
stiffener	加强筋	加強材,防撓材
still water bending moment	静水弯矩	靜水彎矩
stinger	托管架	艉托架（布管船）
stirling engine power plant	热气机动力装置	史特靈引擎動力設備
stock anchor	有杆锚	有桿錨
stockless anchor	无杆锚	無桿錨,山字錨
stopper	阻索器	停止器
stopping test	停船试验	停船性能試驗
storm rails	风暴扶手	風暴扶手
storm valve	防浪阀	止浪閥
stowage	积载	裝載
stowage factor	积载因数	積載因數
stowage plan	集装箱积载图	貨物裝載圖
straight polarity	正接	正極性
straightener (= straightening machine)	矫直机	矯直器
straightening machine	矫直机	矯直器
strain ageing	形变时效	應變老化
strake	列板	列板,板列
stray-current corrosion	杂散电流腐蚀	迷走電流腐蝕
streamline rudder	流线型舵	流線型舵
strength deck	强力甲板	強度甲板
stress	应力	應力
stress concentration	应力集中	應力集中
stress corrosion	应力腐蚀	應力腐蝕
stress corrosion cracking	应力腐蚀断裂	應力腐蝕龜裂
stress relief heat treatment (= stress relieving)	去应力退火	應力消除熱處理
stress relieving	去应力退火	應力消除熱處理
stripping main line	扫舱总管	收艙總管
stripping pump	扫舱泵	殘油泵,收艙泵
structure borne noise	结构噪声	結構噪音
strut	撑材	支桿,支柱
strut bearing	艉轴架轴承	艉軸支架軸承
stud chain	有挡锚链	日字鏈
stud welding	螺柱焊	柱焊,嵌柱焊接
stuffing box	舵杆填料函	舵填料函

英　文　名	祖国大陆名	台湾地区名
subdivision draft	分舱吃水	艙區劃分吃水
subdivision loadline	分舱载重线	艙區劃分載重線
submarine	潜[水]艇	潛[水]艇,潛艦
submarine cable	海底电缆	海底電纜
submarine chaser	猎潜艇	驅潛艇
submarine rescue ship	潜艇救生船	潛艇救難艦
submarine rescue vessel（=submarine rescue ship）	潜艇救生船	潛艇救難艦
submerged arc welding	埋弧焊	潛弧焊,潛溶焊
submerged endurance	水下逗留时间	水下續航力
submerged lamp	水下作业灯	水下作業燈
submerged melt arc welding（=submerged arc welding）	埋弧焊	潛弧焊,潛溶焊
submersible	潜水器	潛水器
submersible drilling platform（=submersible drilling unit）	坐底式钻井平台	坐底式鑽油平台
submersible drilling unit	坐底式钻井平台	坐底式鑽油平台
submersible vehicle（=submersible）	潜水器	潛水載具
subsea production system	水下生产系统	海底石油生産系統
suction dredger	吸扬挖泥船	吸管式挖泥船
suction head	吸头	吸引水頭
suction pipe	吸泥管	吸入管
suction stroke	进气行程	吸氣衝程,吸入衝程
Suez canal search light	苏伊士运河探照灯	蘇彝士運河探照燈
Suez Canal Special Tonnage Certificate	苏伊士运河专用吨位证书	蘇彝士運河噸位證書
Suez canal tonnage	苏伊士运河吨位	蘇彝士運河噸位
sunken mine	沉雷（水雷）	沉雷
super-cavitating propeller	超空化螺旋桨	超空化螺槳
supercharged boiler	增压锅炉	增壓鍋爐
supercharged engine	增压内燃机	增壓引擎
supercooling（=undercooling）	过冷	過[度]冷[卻]
superheater	过热器	過熱器
super high pressure boiler	超高压锅炉	超高壓鍋爐
superstructure	上层建筑	上層建築,船樓建築
superstructure deck	上层建筑甲板	船樓甲板
supply ship（=tender）	供应船	補給船
surface blow off valve（=surface blow-	上排污阀	液面吹泄閥

英　文　名	祖国大陆名	台湾地区名
down valve)		
surface blowdown valve	上排污阀	液面吹泄閥
surface ship	水面舰船	水面船
surface condenser	表面式凝汽器	表面冷凝器
surface force	表面力	表面力
surface ignition	热面点火	表面點火
surface pretreatment	表面预处理	表面預處理
surface roughness	表面粗糙度	表面粗糙度
surface speed	水上航速	水面速[率]
surface treatment	表面处理	表面處理
surface wave	表面波	表面波
surfacing	①上浮 ②堆焊	①上浮 ②堆焊,堆焊接
surge	喘振	波振
surge arrester	避雷器	避雷器
surge line	喘振线	顫動線,喘振線
surge tank	缓冲柜	緩衝櫃
surging	纵荡	縱移
survey ship (=hydrographic survey vessel)	航道测量船	測量船
survivability	生存能力	存活性
swash bulkhead	制荡舱壁	制水艙壁
swash plate	①制荡板 ②制流板	制水板
SWATH (=small waterplane area twin hull)	小水线面双体船	小水線面雙體船
sway (=swaying)	横荡	橫移
swaying	横荡	橫移
sweep	①扫描 ②扫雷具	①掃描 ②掃雷具
swell compensator	涌浪补偿器	波浪補償器
swim-out (=swim-out discharge)	自航发射	[魚雷]自滑發射
swim-out discharge	自航发射	[魚雷]自滑發射
swirler	旋流器	旋流器,回旋式噴
switch	开关	開關
swivel	锚链转环	轉環
swivel piece	锚端链节	轉環
SWL (=safe working load)	吊杆安全工作负荷	安全工作負荷,用負荷
symmetrical flooding	对称浸水	
synchronous generator	同步发电机	同步發電機

英　文　名	祖国大陆名	台湾地区名
synchronous machine	同步电机	同步機
synchronous motor	同步电动机	同步馬達
synchronous speed	同步转速	同步轉速
syren（＝air horn）	气笛	汽笛,氣笛;號笛

T

英　文　名	祖国大陆名	台湾地区名
table of offsets	型值表	船線座標表
tachometer	转速记录器	轉速計
tack welding	定位焊	定位點焊
tactical diameter（＝steady turning diameter）	回转直径	迴旋直徑
T-aerial（＝T-antenna）	T 型天线	T 型天線
take-home engine unit	应急航行机组	
tandem dock	串联式船坞	
tandem propeller	串列螺旋桨	重疊螺槳
tanker	液货船	液货船
T-antenna	T 型天线	T 型天線
tare mass	自重	空重,皮重
tare weight（＝tare mass）	自重	空重,皮重
target craft	靶船	靶船
target ship（＝target craft）	靶船	靶船
tarpaulin	防水盖布	油帆布
Taylor's advance coefficient（＝Taylor's diameter constant）	直径系数	直徑因數（螺槳）
Taylor's diameter constant	直径系数	直徑因數（螺槳）
T-bar（＝T-section）	T 型钢	T 型材
technology of hull construction	船体建造工艺	
temper color	回火色	回火色
template	样板	型板,樣板,模板
tender	供应船	補給船
tensile strength	抗拉强度	拉伸強度
tensioner	张紧器	張力器
tension leg platform（TLP）	张力腿平台	張力腳式鑽油台
terminal box	终端盒	接線盒
test specimen（＝specimen）	试件,试样	試件,試樣
TEU（＝twenty-feet equivalent units）	换算箱	二十英尺貨櫃當量

英　文　名	祖国大陆名	台湾地区名
thermal detector	感温式探测器	溫度探測器
thermal efficiency	热效率	熱效率
thermal fatigue	热疲劳	熱疲勞
thermal load	热负荷	熱負載
thermal protective aid	保温用具	保溫具
thermal resistance	热阻	熱阻
thermal shock	热冲击	熱衝擊
thermal stresses	热应力	熱應力
thermit welding	热剂焊	發熱焊接,鋁熱焊接
thickness gauge	测厚仪	厚度規
thimble	套环	牛眼圈,纜索嵌環
three-dimensional unit	立体分段	
throat depth	喉深	喉深（焊接）
throttle governing	节流调节	節流調速
through bracket	贯通肘板	全通腋板,貫通托架
thrust bearing	推力轴轴承	推力[軸]承,止推[軸]承
thrust block（=thrust bearing）	推力轴轴承	推力[軸]承,止推[軸]承
thrust coefficient	推力系数	推力係數
thrust collar	推力盘,推力环	推力[軸]環
thrust deduction factor	推力减额因数	推[力]減[少]因數
thrust meter	推力仪	推力計
thrust pad	推力块	推力墊,止推墊
thrust shaft	推力轴	推力軸
thruster	侧推器	推力裝置
tie insert	金属扣合	金屬扣合
TIG welding	钨极惰性气体保护焊	氣體遮蔽鎢弧焊
tiller	舵柄	舵柄
tiller tie-bar	舵柄连杆	舵柄連桿
timber carrier	运木船	木材運輸船,運木船
time-delay relay	时间继电器	延時繼電器
time lag	时间间隔	時間延遲,時滯
tip cavitation	叶梢空化	葉尖空化
tipping	艉落	艉驟降
T-joint	T形接头	T形接頭,三通管接
TLP（=tension leg platform）	张力腿平台	張力腳式鑽油台
toe crack	焊趾裂纹	趾裂痕（焊接）

英　文　名	祖国大陆名	台湾地区名
toe of weld	焊趾	焊接縫突趾
tolerance of hull construction	船体建造公差	
tonnage	吨位	噸位
tonnage measurement	吨位丈量	噸位丈量
topcoat	面层	表塗層
top dead center	上止点	上死點
topgallant mast	上桅	上桅
topping bracket	千斤座	［桅頂］俯仰滑車座
topping lift	千斤索	俯仰頂索（吊桿）
topping winch	顶索绞车	俯仰絞機（吊桿）
top side tank	顶边舱	［翼］肩艙
torch	焊矩	吹把
torpedo boat	鱼雷快艇	魚雷［快］艇
torpedo launcher for submarines (= torpe-do tube)	潜艇鱼雷发射装置	魚雷發射管
torpedo motor boat (= torpedo boat)	鱼雷快艇	魚雷［快］艇
torpedo recovery ship	捞雷船	撈雷船
torpedo tube	潜艇鱼雷发射装置	魚雷發射管
torque	扭矩	扭矩,轉矩
torque coefficient	转矩系数	轉矩係數
torque meter	转矩仪	轉矩計
torsional vibration	扭转振动	扭轉振動
torsional vibration of shafting	轴系扭转振动	軸系扭轉振動
total displacement	总排水量	總排水量
totally enclosed lifeboat	全封闭救生艇	全圍蔽救生艇
total resistance	总阻力	總阻力
toughness	韧性	韌性,韌度
tourist submersible	水下游览船	水下遊覽船
tow boat (= tug)	拖船	拖船
towing arch	拖曳弓架	拖纜拱架
towing beam	拖缆承架	拖索承梁
towing hook	拖钩	拖［纜］鈎
towing light	拖带灯	拖航燈
towing line	拖缆	拖纜,拖索
towing post	拖桩	拖纜樁
towing speed	拖曳航速	拖曳船速
towing tank	拖曳水池	船模試驗槽
towing vessel (= tug)	拖船	拖船

英　文　名	祖国大陆名	台湾地区名
towing winch	拖缆绞车	拖纜絞機
towrope（=towing line）	拖缆	拖纜,拖索
trade route（=shipping route）	航线	航線
trailer	挂车	拖車
trailing edge	随边	殿緣
trailing suction hopper dredger	耙吸挖泥船	
training ship	训练舰,训练船	訓練船
training vessel（=training ship）	训练舰,训练船	訓練船
transducer	换能器	換能器
transfer	横距	迴旋横距
transfer chamber	过渡舱	過渡艙
transfer platform	输油平台	輸油平台
transient current	瞬态电流	暫態電流
transient mode（=transient state）	瞬态	暫態,過度狀態,瞬時狀態
transient state	瞬态	暫態,過度狀態,瞬時狀態
transient wave test	瞬态波试验	暫態波試驗
transmission efficiency of shafting	轴系效率	軸系效率
transmission gear of shafting	轴系传动装置	軸系傳動裝置
transom	方艉端面	艉橫板,艉橫材
transom beam	艉横梁	艉梁
transom floor	艉肋板	艉肋板
transom stern	方艉	平艉
transport ship	运输船	運輸船
transverse bulkhead	横舱壁	横向艙壁
transverse center of gravity	重心横向坐标	横向重心
transverse forced oscillation device	横向强制摇荡装置	横向強制搖擺裝置
transverse frame system（=transverse framing system）	横骨架式	横肋系統
transverse framing system	横骨架式	横肋系統
transverse metacenter	横稳心	横定傾中心
transverse metacentric height	横重稳距	横向定傾高
transverse sections	横剖面	［横］剖面
transverse stability	横稳性	横向穩度
transverse strength	横向强度	横向強度
transverse wave	横波	横波
trawler	拖网渔船	拖網漁船

英 文 名	祖国大陆名	台湾地区名
trawl gallow	网板架	網板架
trial speed	试航速度	試航船速
triangular plate	三角眼板	三角眼板
trim	纵倾	俯仰
trimaran	三体艇	三［胴］體船
trim by bow	艏倾	艏俯
trim by stern	艉倾	艉俯
trimming	平舱	整平（散装货）
trimming moment	纵倾力矩	俯仰力矩
trimming pump	纵倾平衡泵	俯仰水泵
triplex pump	三联泵	三缸泵
tripod mast	三脚桅	三腳桅
tripping bracket	防倾肘板	防撓肘板
troller（＝trolling boat）	曳绳钓渔船	曳繩釣漁船
trolling boat	曳绳钓渔船	曳繩釣漁船
trolling gurdy	曳绳钓起线机	曳繩釣起繩機
truck	桅冠	桅頂
trunk bulkhead	围壁	圍壁
T-section	T 型钢	T 型材
tube plate	管板	管板
tube shaft（＝screw shaft）	螺旋桨轴,艉轴	艉［管］軸
tubular joint	管结点	管接頭
tubular pillar	管形支柱	圓管柱
tug	拖船	拖船
tugboat（＝tug）	拖船	拖船
tumble home	内倾	船舷內傾
tuning factor	调谐因子	調諧因子
tunnel recess	轴隧艉室	軸道凹部
tunnel stern	隧道艉	隧［道］艉
tunnel top line	隧道顶线	隧［道］頂線
turbine steam seal system	汽封系统	渦輪汽封系統
turbo-blower	汽轮鼓风机	渦輪鼓風機
turbocharger	涡轮增压器	渦輪增壓機
turbo-charging	废气涡轮增压	渦輪增壓
turbulence detector	紊流探测器	紊流探測器
turbulence stimulator	激流装置	激紊裝置
turning circle	回转区域	回旋圈
turning circle radius	回转区域半径	回旋圈半徑

英　文　名	祖国大陆名	台湾地区名
turning path	回转迹线	回旋跡線
turning period	回转周期	回旋周期
turning test	回转试验	回旋試驗
turning trial（＝turning test）	回转试验	回旋試驗
turntable	转台	轉台
turntable of dipper machine	铲斗转盘	鏟斗轉盤
turret moored drilling ship（＝center moored drilling ship）	中心系泊定位钻井船	中心系泊定位鑽探船
tweendeck cargo space（＝tweendeck space）	甲板间舱	甲板間艙
tweendeck frame	甲板间肋骨	甲板間肋骨
tweendeck space	甲板间舱	甲板間艙
twenty-feet equivalent units	换算箱	二十英尺貨櫃當量
twin-hull ship（＝catamaran）	双体船	雙[胴]體船
twin-skeg	双艉鳍	雙艉鰭
twisting	扭曲	扭曲
two boat trawler（＝bull trawler）	双拖渔船	雙拖[網]漁船
two circuit feeding	双路供电	雙路供電,雙重供電
two-direction truss	双向桁架	雙向桁架
two-gate caisson（＝mitre caisson）	人字式坞门	人字式塢門
two-part hull construction	两段造船法	兩船段建造法
two-stroke cycle	二冲程循环	二衝程循環
two-stroke engine	二冲程内燃机	二衝程引擎
type	型式	型,式

U

英　文　名	祖国大陆名	台湾地区名
ultra short wave communication	超短波通信	
umbilical	脐带	[潛水具]供應連系管,脐索
unattended machinery space	无人[值班]机舱	無人[當值]機艙空間
unbalanced rudder	不平衡舵	不平衡舵
unburied pipeline	裸置管道	裸置管路
under bead crack	焊道下裂纹	焊珠底龜裂
undercooling	过冷	過[度]冷[卻]
undercut	咬边	過熔低陷（焊接）
underside handholds	舷部扶手	舷部扶手

英　文　名	祖国大陆名	台湾地区名
underwater acoustic absorption material	水声吸声材料	水下吸音材
underwater acoustic reflection material	水声反声材料	水下反音材
underwater acoustic research ship（ ＝under- derwater acoustic research vessel）	水声调查船	水聲調查船
underwater acoustic research vessel	水声调查船	水聲調查船
underwater acoustic transmission material	水声透声材料	水下透音材
underwater adhesion	水下黏合	水下黏合
underwater adhesive	水下胶黏剂	水下膠黏劑
underwater cutting	水下切割	水中切割
underwater explosion tank	水下爆炸试验水池	水下爆炸試驗水槽
underwater habitat	水下居住舱	水下居住艙
underwater monitoring station	水下监听站	水下監測站
underwater operating machine	水下作业机械	水下作業機械
underwater ship	全潜船	全潛船
underwater sightseeing-boat（ ＝tourist submersible）	水下游览船	水下遊覽船
underwater welding	水下焊接	水中焊接
underwater working station	水下作业站	水下工作站
underway replenishment ship	航行补给船	
undock（ ＝undocking）	出坞	出塢
undocking	出坞	出塢
uniflow scavenging	直流扫力	單流驅氣
union joint	螺纹接头连接	活管套接頭
union purchase system	双杆吊货装置	雙吊桿作業系統
unit assembling	单元组装	單元組合,小組合
unit outfitting	单元舾装	
universal chock	转动导缆孔	萬向導纜孔
universal coupling	万向联轴器	萬向聯結器
unmanned submersible	无人潜水器	無人潛水器
unsteady cavities	不定常空泡	不穩定空泡
unsymmetrical flooding	不对称浸水	不對稱浸水
untethered remotely operated vehicle	无缆遥控潜水器	無纜遙控潛水器
upper bearer（ ＝rudder carrier）	上舵承	舵承上部,上舵承
upper deck	上甲板	上甲板
uprighting analysis	扶正分析	扶正分析
upsetting	镦粗	鐓鍛,鍛粗
U-section	U 型剖面	U 型剖面
useful life	使用寿命	有效壽命

英 文 名	祖国大陆名	台湾地区名
utility boat (= work boat)	工作艇	工作船,工作艇
utilization coefficient for sacrificial anode	牺牲阳极利用效率	犧牲陽極利用效率

V

英 文 名	祖国大陆名	台湾地区名
vacuum	真空度	真空度
vacuum breaker	真空破坏器	真空破除器
vacuum condenser	真空冷凝器	真空冷凝器
vacuum test	真空试验	真空試驗
valve lapping	阀面研磨	閥面研磨
valve lift	气门升程	閥升程
valve timing	配气定时	閥動定時
vapor and water mixing heater	混合式水加热器	混合式水汽加熱器
vaporizing oil range	汽化油灶	汽化油竈
variable ballast tank	可调压载水舱	可變壓載艙
variable load	可变载荷	變動負荷,變動負載
variable pitch	变螺距	可變節距,可變螺距
variable speed governor	全程式调速器	可變調速器
V-drive	V 型传动	V 形驅動
vehicle	载体	載具
vehicle deck (= wagon deck)	车辆甲板	車輛甲板
vehicle hold	车辆舱	車輛艙
velocimeter (= velocity meter)	流速计	速度計,測速器
velocity coefficient	速度系数	速度係數
velocity meter	流速计	速度計,測速器
velocity stage	速度级	速度級
vertical boiler	立式锅炉	立式鍋爐
vertical bow	直立型艏	直立型艏
vertical center of gravity	重心垂向坐标	垂向重心
vertical engine	立式内燃机	立式引擎
vertical flexural vibration	垂向弯曲振动	垂向彎曲振動
vertical girder	竖桁	豎桁
vertical ladder	直梯	直立梯
vertical prismatic coefficient	垂向棱形系数	垂向棱塊係數
vertical template for hull assembly	船台标杆线	船台標竿線
vertical-type resistance dynamometer	垂直式阻力动力仪	垂直式阻力動力計
very large crude oil carrier (VLCC)	巨型油轮	極大型原油輪,巨型油

英　文　名	祖国大陆名	台湾地区名
		輪
very long wave communication	甚长波通信	
vessel（＝ship）	船［舶］	船，艦
vibration	振动	振動
vibration exciter	激振机	激振機
vibration severity	振动烈度	振動烈度
viewport	观察窗	檢查孔，觀察孔
virtual mass	实效重量	虛質量
viscosity	黏度	黏度
viscous pressure resistance	黏压阻力	黏性壓差阻力
viscous resistance	黏性阻力	黏性阻力
visual signal	光信号	視覺信號
VLCC（＝very large crude oil carrier）	巨型油轮	極大型原油輪，巨型油輪
void space（＝air tank）	浮力舱	空艙
Voith-Schneider propeller（＝cycloidal propeller）	平旋推进器 槳	擺線推進器，垂直翼螺槳
voltage reducing device	［空载］电压降低装置	減壓裝置
voyage	航次	航次

W

英　文　名	祖国大陆名	台湾地区名
wagon deck	车辆甲板	車輛甲板
wake	伴流	伴流，跡流
wake-adapted propeller	适伴流螺旋桨	適跡［流］螺槳
wake coefficient（＝wake fraction）	伴流分数	跡流因數，跡流係數
wake factor	伴流因数	跡流因數，跡流係數
wake fraction	伴流分数	跡流因數，跡流係數
wake measurement	伴流测量	跡流量測
wake simulation	伴流模拟	跡流模擬
wake survey（＝wake measurement）	伴流测量	跡流量測
Ward-Leonard system（＝generator-motor system）	发电机电动机系统	列氏電動操作系統
warp block	曳纲滑轮	曳網滑輪
warping capstan	引船绞盘	捲索絞盤
warping end	绞缆筒	捲索鼓
warping winch	绞缆绞车	捲索絞機

英　文　名	祖国大陆名	台湾地区名
warship（＝naval ship）	舰艇	軍船
wash-back（＝set-back）	翘度	翹度
washing	洗舱	洗艙
washing boiler	洗炉	洗爐
washing equipment	清洗装置	清洗裝置
washing test	清洗试验	清洗試驗
waste heat boiler	余热锅炉	廢熱鍋爐
water ballast	水压载	水壓載
water cooling	水冷	水冷卻
water ejector	射水抽水器	射水抽水器
water filling test	灌水试验	注水試驗
water jet［propelled］boat	喷水推进船	噴水推進船，噴射推進船
water jet propulsion	喷水推进	噴水推進
water-jet propulsion gas turbine	喷水推进燃气轮机	噴水推進燃氣渦輪機
waterjet propulsor	喷水推进器	噴水推進器
water level	水位	水位
water level elevation	水面高程	水面高程
waterline	水线	水線
waterline length	水线长	水線長度
waterline zone corrosion	水线区腐蚀	水線區腐蝕
water plane	水线面	水線面
water pouring test	淋水试验	淋水試驗
water resistance	水阻	水阻力
water seal scupper	水封式漏水口	水封式排水口
water-softening plant	给水软化装置	給水軟化裝置
watertight aircase	空气箱	水密空氣箱
watertight bulkhead	水密舱壁	水密艙壁
watertight door	水密门	水密門
watertight floor	水密肋板	水密肋板
watertight subdivision	水密分舱	水密艙區劃分
water tube boiler	水管锅炉	水管鍋爐
wave	波	波
wave breaking resistance	破波阻力	碎波阻力
wave clearance	峰隙	距波高度
wave form	波形	波形
wave front	波前	波前［進面］
wave generator	造波机	造波機

英 文 名	祖 国 大 陆 名	台 湾 地 区 名
wave length	波长	波長
wave load	波浪载荷	波浪負荷
wave loop（=antinode）	波腹	波腹
wave maker（=wave generator）	造波机	造波機
wave making damping	兴波阻尼	興波阻尼
wave making resistance	兴波阻力	興波阻力
wave pattern resistance	波型阻力	波型阻力
wave spectrum	波浪谱	波譜
wave train	波列	波列
way end pressure	滑道末端压力	軌端壓力（下水）
wear resistance	耐磨性	耐磨耗性
weather ship	天气船	氣象觀測船
weather vessel（=weather ship）	天气船	氣象觀測船
weather tight door	风雨密门	風雨密門
weathertightness	风雨密性	風雨密性
web	腹板	腹板,大肋骨
web beam	强横梁	強力梁,大梁
web frame	强肋骨	大肋骨
wedge	斜楔	尖劈,楔
weight curve	重量曲线	重量曲線
weight replacing system	重量代换系统	重量代換系統
weight ton	重量吨	重量噸
weld	焊缝	焊道
weldability	焊接性	可焊性
weld corrosion（=weld decay）	焊区腐蚀	焊道[晶間]腐蝕
weld decay	焊区腐蚀	焊道[晶間]腐蝕
welding	焊接	焊接[法]
welding arc	焊接电弧	焊接電弧
welding by both side	双面焊	雙面焊接接合
welding condition	焊接工艺参数	焊接條件
welding current	焊接电流	焊接電流
welding flux	焊剂	焊藥,助焊劑
welding joint	[焊接]接头	焊接接頭
welding operation	焊接操作	焊接操作
welding parameter（=welding condition）	焊接工艺参数	焊接條件
welding power source	焊接电源	焊接電源
welding rod extrusion press	焊条压涂机	焊條塗藥機
welding sequence	焊接顺序	焊接順序

英 文 名	祖 国 大 陆 名	台 湾 地 区 名
welding speed	焊接速度	焊接速度
welding stress	焊接应力	焊接應力
welding technique	焊接技术	焊接技術
welding with backing	衬垫焊	襯墊焊法
weld length	焊缝长度	焊道長度
weldment	焊件	焊件
weld metal	焊缝金属	焊接金屬
weld metal area	焊缝区	焊接金屬面積
weld spacing	焊点距	焊道間距
weld time	焊接通电时间	焊接時間
wellhead platform	井口平台	井口平台
wet and dry pipe valve installation	干湿两用阀装置	乾濕兩用閥裝置
wet chamber	湿舱	濕艙
wet on wet	湿碰湿	濕式積層
wet sand blasting	湿喷砂除锈	濕噴砂除銹
wet strength	湿态强度	濕態強度
wet submersible	湿式潜水器	濕式潛水器
wetted surface	湿面积	浸水面
whale factory ship	捕鲸母船	鯨加工船
whaler	捕鲸船	捕鯨船
wharf	码头	碼頭
wheel house	驾驶室	駕駛室,[操]舵房
whipping	冲荡	抖動
whistle and siren control system	号笛控制装置	號笛控制器
whistle controller (= whistle and siren control system)	号笛控制装置	號笛控制器
wildcat (= cable lifter)	锚链轮	鏈輪
winch	绞车	絞車,絞機
winch platform	起货机平台	絞機台
windlass	起锚机	錨機
wind scooper	导风罩	導風罩
wind wave and current tank	风、浪、流试验水池	風、浪、流試驗水槽
wing tank	边舱	翼櫃,翼艙
wing wall	坞墙	浮塢牆
wire drive feed unit	送丝装置	供[焊]線裝置
wire feeder	送丝机构	供線機
wooden ship (= wooden vessel)	木船	木船
wooden vessel	木船	木船

英　文　名	祖国大陆名	台湾地区名
work boat	工作艇	工作船,工作艇
working medium volume	气缸容积	氣缸容積
working ship	工程船	工作船,作業船
working stroke	工作行程	工作衝程
workpiece	工件	工作件
work ship（＝working ship）	工程船	工作船,作業船
work shop	机修间	工場
wrinkles	起皱	皺紋
wrong operation alarm	误操作报警器	誤操作警報［器］

X

英　文　名	祖国大陆名	台湾地区名
X-joint	X 型结点	X 型接頭
X-radiation	X 射线	X 射線
X-ray（＝X-radiation）	X 射线	X 射線

Y

英　文　名	祖国大陆名	台湾地区名
Y joint	Y 型结点	Y 型接頭
yacht（＝pleasure craft）	游艇	遊艇
yard	桅横杆	帆桁
yawing	艏摇	［艏艉］平擺
yoke	横舵柄	横舵柄

Z

英　文　名	祖国大陆名	台湾地区名
Z-drive	Z 型传动	Z 型驅動
zero trim（＝floating on even keel）	正浮	縱平浮
zigzag test	Z 形［操纵］试验	Z 形［操縱］試驗.蛇航［操縱］試驗
Z-peller propulsion	Z 型推进	Z 型推進